Calculus I

A Guided Inquiry

PRELIMINARY EDITION

a process oriented guided inquiry learning course

Andrei Straumanis — *The POGIL Project*

Catherine Bénéteau — *University of South Florida*

Zdeňka Guadarrama — *Rockhurst University*

Jill E. Guerra — *University of Arkansas Fort Smith*

Laurie Lenz — *Marymount University*

THE POGIL PROJECT, LANCASTER, PA

POGIL Project
Director: Richard Moog
Associate Director: Marcy Dubroff
Publication Liaison: Sarah Rathmell

ISBN-13: 978-1-118-87748-7
ISBN: 1-118-87748-9

for calculus students and teachers

Acknowledgements

Special thanks goes to our students. The process of revising and improving these activities was greatly enhanced by the feedback that came from watching students work the activities during class.

Thanks also to the *beta* testers and their students who provided feedback on the *beta* version. These were Lisa DeCastro of the University of South Florida, Garrett Gregor of Clark College, Kathy Pinzon of Georgia Gwinnett College, and Rachel Weir of Allegheny College.

This book would not have been possible without the POGIL Project and its growing number of participants. Huge credit goes to Rick Moog, the founder and Director of the POGIL Project, whose contagious enthusiasm for guided inquiry has inspired many to embark on this path.

Funding was provided by the National Science Foundation grant numbers 1123061, 1122747, 1122757, 1122600, 1122695, and initial funds were provided by the United States Department of Education's Fund for Improvement of Post-secondary Education (FIPSE) under grant number P116B060026.

Forward on Process Oriented Guided Inquiry Learning

by Rick Moog, Jim Spencer and John Farrell, authors of *Chemistry: A Guided Inquiry* and two volumes of *Physical Chemistry: A Guided Inquiry.*

These guided activities were written because much research has shown that more learning takes place when the student is actively engaged and when ideas and concepts are developed by the student, rather than being presented by an *authority*—a textbook or an instructor.[1] The activities presented here are structured so that information is presented to the reader in some form (an equation, table, graph, figure, written prose, etc.) followed by a series of *Construct-Your-Understanding Questions* that lead the student to the development of a particular concept or idea. Learning follows the scientific process as much as possible throughout. Students are often asked to construct a concept based on the model that has been developed up to that point, and then further data or information is provided to help refine the concept. In this way, students simultaneously learn course content and key process skills that constitute mathematical and scientific thought and exploration.[2,3]

---

[1] Johnson, D. W.; Johnson, R. T. *Cooperative Learning and Achievement.* In Sharon, S. (Ed.), *Cooperative Learning: Theory and Research*, pp 23-37, New York: Praeger, 1996.

[2] *How People Learn: Brain, Mind, Experience, and School*; Bransford, J.D., Brown, A.L., Cocking, R.R., editors; National Research Council, National Academy Press, Washington, D.C. 1999.

[3] Farrell, J.J., Moog, R.S., Spencer, J.N., "A Guided Inquiry General Chemistry Course." *J. Chem. Educ.*, 1999. 76: 570-574.

Comments from Faculty about POGIL

In adopting a POGIL format in a large classroom, my day-to-day preparations were comparable in intensity and duration to the time I spent preparing for traditional lectures. In my years as a college educator, I have not seen anything as pedagogically powerful as a POGIL class. I believe the future of college 'teaching' lies in this type of 'learning.'

The guided inquiry helps me [the professor] think more like a student and it helps me cover material more efficiently.

The students love the interactive nature of class time. I become better acquainted with each of my students, which enables me to tailor my teaching to maximize each student's learning. It has transformed the way I teach.

Watching my students engaged in their groups during class has convinced me that I made the right decision to change to POGIL after twenty years of brilliant lecturing. The activities really do help the students learn much better than my lectures ever did.

Comments from Students about POGIL

I didn't get tired during class because I was constantly thinking and working instead of in a lecture class where I just listen and get easily tired.

The act of explaining the concept forced me to clarify the concept in my own head.

Class time was actually learning time, not just directly-from-ear-to-paper-and-bypass-brain-writing-down time. Learning the material over the whole term is far easier than not "really" learning it until studying for the tests.

Overall, I was far less stressed than many of my friends who took the lecture class. They basically struggled through everything on their own.

Group work has helped me find motivation for studying.

It was hugely beneficial to be able to discuss through ideas as we were learning them; this way it was easy to immediately identify problem areas and work them out before going on.

The method of having us work through the material for ourselves— as opposed to being told the information and trying to absorb it—makes it seem natural or intuitive. This makes it very nice for learning new material because then we can reason it out from what we already know.

I felt like I was actually learning the information as I received it, not just filing it for later use. The format helped me retain much more material than I have ever been able to in a lecture class, and the small, group atmosphere allowed me to feel much more comfortable asking questions of both other students and the professor.

How to Use this Book (for Instructors)

Process Oriented Guided Inquiry Learning (POGIL) is used to teach tens of thousands of science and mathematics students each semester. These materials are flexible and can be used in a variety of settings, but during most POGIL sessions…

- **Students work in teams of three or four** to answer the *Construct-Your-Understanding Questions*. (This group work is in the place of formal lecture.) These questions are not homework questions; they are carefully designed to guide student groups toward discovery of a concept.
- **Group work does not mean group grading**. Individual accountability is essential for motivating each student to take advantage of the resources of his or her group and the POGIL classroom. So, as in a traditional classroom, quizzes and exams should be taken individually.
- **The instructor serves as the facilitator of learning,** not as the primary source of information. Though effective facilitators spend significant class time observing student group work, they are not on the sidelines. During each class they ask and answer questions, lead whole-class discussions, and deliver just-in-time mini-lectures (usually, no more than five minutes long).
- **Especially at the start, students will ask you: "Is our answer right?"** Answering this question reinforces students' belief that their job is to memorize information presented or endorsed by an authority (you). Remind them that their job is to *construct a valid understanding of the underlying concepts*. Telling them the right answer can short-circuit this, and bring their processing of the ideas to a premature end. Ask how you can help. Usually students can rephrase their question to highlight the source of their confusion.
- **Many instructors choose to assign group roles**, and rotate them frequently. Some roles to try:
    - **Manager** – In charge of time management, promotion of equal participation, and ensuring that no group member is falling behind or dashing ahead. The manager is often charged with posing all group questions to the instructor.
    - **Presenter** – Presents the group's work during whole-class discussions.
    - **Reader** – Reads aloud each question for the group. This role can seem awkward at first, but we have found no better way to promote group interaction among students who are new to the POGIL method.
- **At the start of each class, give a 1-question (paper or clicker) quiz** on a concept from the previous activity. This encourages steady effort and, since the activities build on one another, it ensures that students come to each class prepared to contribute to and learn from group work.
- **The activities in this book are designed to be a student's <u>first</u> introduction to a topic.** A very short (3-minute) review of the previous activity can be helpful when delivered at the start of the *next* class (right after the quiz). If you are employing a flipped classroom model or other structure that includes some formal lecture, make sure that any lecture on a topic comes *after* students have had the opportunity to discover that topic by completing the activity.
- **This workbook can be used with any textbook**. Most POGIL instructors assign homework and reading from a textbook. Note the important change that students should be asked to read the relevant sections of the textbook *after* they have completed the activity on that topic.

For more information, visit www.guidedinquiry.org, or contact the POGIL Project at www.pogil.org. Another good resource is the instructor-only Google group moderated by the authors of these materials, and dedicated to POGIL Mathematics. Instructors can find it by searching "**POGIL Mathematics**" at www.google.com/forum.

How to Use this Book (for Students)

Many students find it useful to read this page at the start of the course, and again two weeks into the course.

This book is designed to make calculus more enjoyable and less intimidating, but without sacrificing depth. Too many students memorize example problems or formulas, only to forget them after the exam. This workbook guides you toward a deep understanding so you learn more, retain it longer, and do better in this course, subsequent courses and on standardized exams. These positive results have been demonstrated using the Process Oriented Guided Inquiry Learning (POGIL) method in many different courses, in many different settings.

This book is designed to be used by student groups, during class. For each activity, read the **Model** then work with your group to answer the *Construct-Your-Understanding Questions* that follow. Stay together! For each question, be sure to compare answers within the group before moving onto the next question. Even if your instructor does not **assign the role of Reader**, try assigning one person to read each question aloud. This can help keep the group together and lead to a better understanding of the material.

If you are unsure of an answer, even after checking with your group, some good strategies are to i) read the next question, ii) ask a nearby group, or iii) pose a question to the instructor, but try to avoid asking the instructor: "***Is our answer right?***" Instead, explain why you are confused, or ask a question that gets at the source of your confusion.

It is a good idea to **assign a group Manager**, and rotate this role frequently. Often, the manager is the only person in the group who is allowed to ask a question of the instructor. This rule is *not* put in place to give the manager an unfair advantage. The purpose is to get the group to talk about each question. This often results in the group answering the question themselves, or in a better formulated question.

If you have no manager, then it is your collective responsibility to make sure everyone in the group participates, and to manage your time so that you finish all the *Construct-Your-Understanding Questions* before the end of class. Before the next class, finish all assigned parts of the activity including homework and reading from a textbook.

Advice to Students from Past Students

Don't let yourself take the course lightly just because class is fun and relaxed (and goes by fast!) Do the homework and reading.

Give yourself some time to settle into group learning. Lots of us did not think we would like it or that it would work. It does.

Don't fall behind. Playing catch up is not fun. Don't be afraid to ask questions and argue in your group. That is the way learning is done in this class.

Find a study group ASAP and meet regularly [outside of class] every week. I wish I had done this sooner.

Contents

Functions

Limits

Derivatives

Differentiation Techniques

Differentiation Applications

Integration

Functions 1: Review of Functions

Model 1: Input-Output Machines

Each machine below reads an input, and then produces a corresponding output, forming a unique pair in which the input and the output are related to one another by some rule that the machine uses to operate.

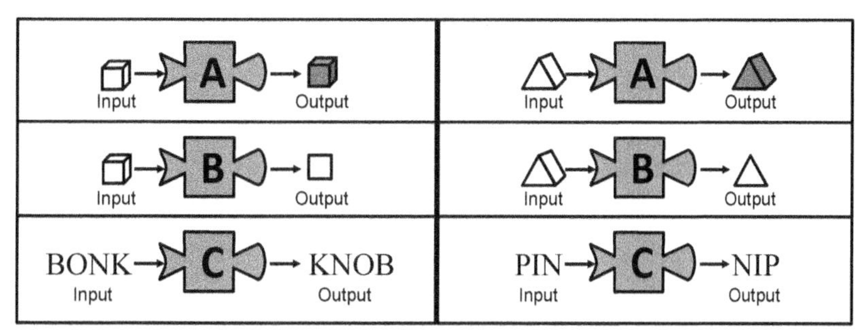

It may be specified that a particular machine can only accept certain types of inputs. For example, we may say Machine A will only accept a white object, or Machine C will only accept a string of letters.

Construct Your Understanding Questions (to do in class)

1. Describe in words the rule, that is, how the output is related to the input for...
   a. Machine A in Model 1

   b. Machine B in Model 1

   c. Machine C in Model 1

2. Draw or write the output of...
   a. Machine B when the input is the cylinder shown at right. 

   b. Machine C when the input is POTS.

3. Describe in words how the output is related to the input for Machine f.

10 → f → 29	1 → f → 2
6 → f → 17	51 → f → 152
2 → f → 5	0 → f → -1

4. What is the output of Machine f if the input is 3?

5. Which is the best representation of the Output of Machine f in terms of the input? Note that in each expression <u>the input is represented by the letter x</u>.

   Output of Machine $f =$

   a. $2x+1$
   b. $3(x-1)$
   c. $3x-1$
   d. x^2+1
   e. x^2-1

6. (Check your work) All but one of the following pairs of inputs and outputs are from Machine f. Cross out the <u>one</u> pair that is not from Machine f.

Input	Output
-1	-4
0.5	0.5
1.5	1.5
5	14
12	35
100	299

7. Inputs and outputs of machines like f are very often plotted on a graph. By convention, the inputs are listed along the horizontal axis, and the outputs along the vertical axis. The graphs of two new machines (Machine g and Machine h) are shown below.

   a. Which machine (g <u>or</u> h) gives an output for any input (x) between -7 and 7?

   b. The horizontal axis is called the x-axis because x is the letter most often used for inputs. Label each horizontal axis of the graphs of g and h with the letter "x".

   c. By convention, the vertical axis (outputs) is called the y-axis. Label each of these "y".

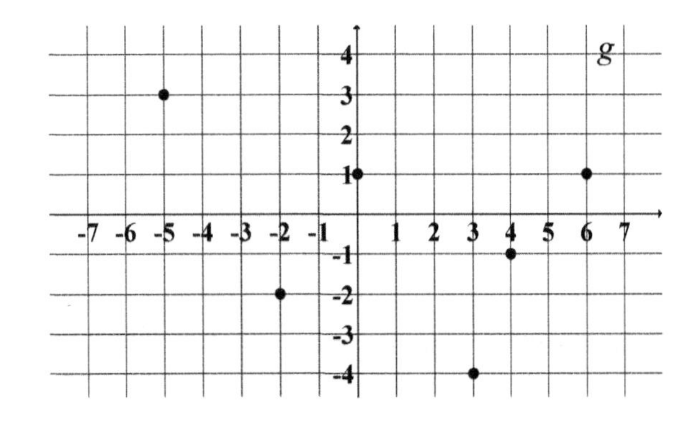

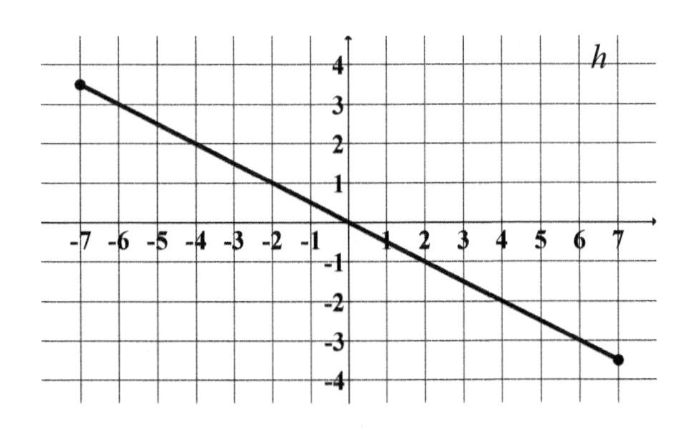

   d. The symbol for the output of Machine g with an input of x is "$g(x)$". What is…

      i. $g(3) =$

      ii. $g(0) =$

      iii. $g(-5) =$

      iv. $h(-4) =$

      v. $h(0) =$

      vi. $h(2) =$

e. Cross out the <u>one</u> statement about Machines g and h that is FALSE.
Recall that the coordinates of a point on a graph are written in the form (x, y).

i. $g(0) = g(6)$

ii. The point $(6, g(6))$ is on the graph of g.

iii. On the graph of g, $y = g(x)$; and on the graph of h, $y = h(x)$.

iv. $(x, g(x))$ is on the graph of g for all values of x in $[-7,7]$.

v. $(x, h(x))$ is on the graph of h for all values of x in $[-7,7]$

8. (Check your work) Statement iii in the previous question is true. (Check that this matches your answer.) Students sometimes try to read too much into this. All it says is: for a given machine (e.g. Machine g) the **output** can be represented by the symbol y or the symbol $g(x)$. Both notations will be used. To remind you of this, on the previous page, add "$y = g(x)$" to the graph of g, and "$y = h(x)$" to the graph of h.

9. Which graph below represents the inputs and outputs of Machine f from Questions 3-6? Label it $f(x) = 3x - 1$.

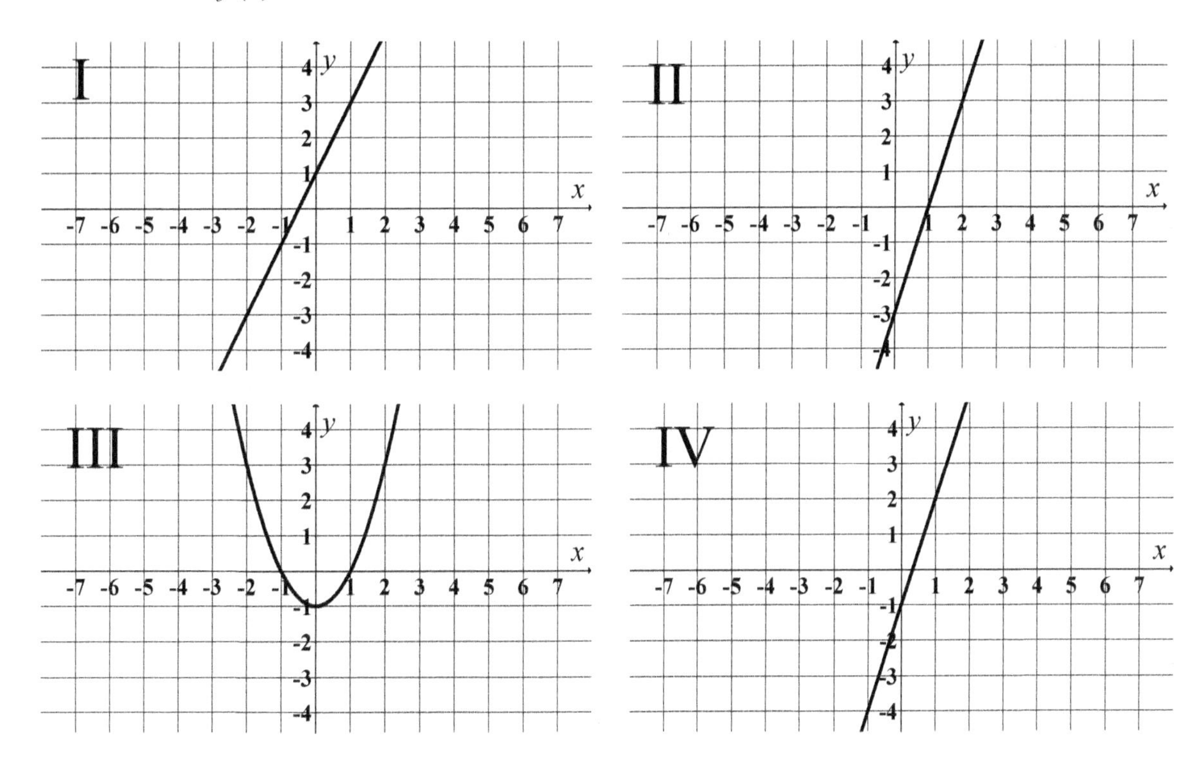

10. Each graph in the previous question is associated with an expression of x found in Question 5. Write the correct expression on each of the other three graphs.

11. Describe the rule or rules associated with each of the following machines.

Machine D	Machine E
D	E
D	E
D	E
D	E

12. Draw the outputs if the shape shown at right is used as an input in Machine D .... and as an input in Machine E.

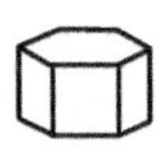

13. According to Summary Box F1.1, which is a **function**: Machine D, Machine E, or both? Circle one, and explain your reasoning.

Summary Box F1.1: Definition of a Function

A **function** is a rule (or machine) that assigns exactly one output to each input.

Special notation is used to represent a mathematical machine that is a **function**.

For example, Machine f from this activity is represented in symbols as $f(x)=3x-1$.

- The input is represented by x .
- The output is represented by $f(x)$, which is read "f of x" .

The graph of a function f is the set of all ordered pairs (x, y) such that $y=f(x)$.

Extend Your Understanding Questions (to do in or out of class)

14. Which of the following is a representation of a function according to the definition in Summary Box F1.1? Explain your reasoning.

a.

b.

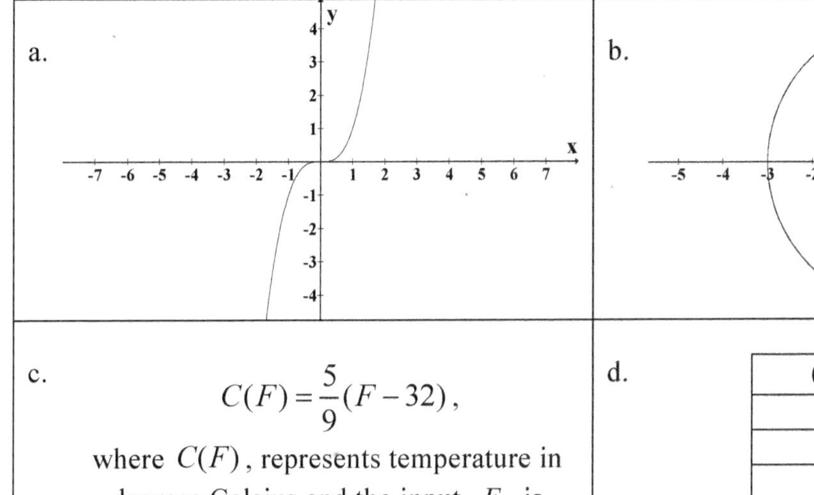

c.

$$C(F) = \frac{5}{9}(F - 32),$$

where $C(F)$, represents temperature in degrees Celsius and the input, F, is temperature in degrees Fahrenheit.

d.

(input)	(output)
9	3
9	-3
25	5
25	-5

e.

Zoo Admission Prices	
five & under	$3
6-12	$6
13-64	$12
65 & over	$6

f.

Confusing Admission Prices	
Infants (2 & under)	free
Kids 2-18	$5
Adults (18 & over)	$10

15. (Check your work) Consider the two different sets of admission prices in the previous question.
    a. Does one or the other make more sense to you? Explain why and cite any problems or ambiguities that you see in either set.

    b. Which set of admission prices, if either, is a function? Explain your reasoning.

    c. If one of these sets of admission prices is <u>not</u> a function, suggest a change that would make it a function.

    d. Comment on whether your proposed change would clear up any ambiguities you cited in part a of this question.

16. Complete Summary Box F1.2 by writing in the word **input** or **output** in both blanks, then cross out the wrong ending to the sentence.

Summary Box F1.2: The Multiple ________________ Test

The test to check if a set of inputs and outputs is a function is called the **Multiple __________ Test.**

It is called this because a collection of inputs and outputs does NOT represent a **function** if and only if it is... [cross out the incorrect ending]

- **possible to find two or more outputs associated with one given input.**

or

- **possible to find two or more inputs associated with one given output**

17. Write a sentence explaining how the Multiple Output Test verifies the conclusion that three of the six situations in Question 14 do not represent a function.

18. A student has drawn the bold vertical line shown at right on her paper and claims that this line shows that the graph is not the graph of a function.
    a. Do you agree?

    b. Explain how such a vertical line might represent an application of the Multiple Output Test to this graph.

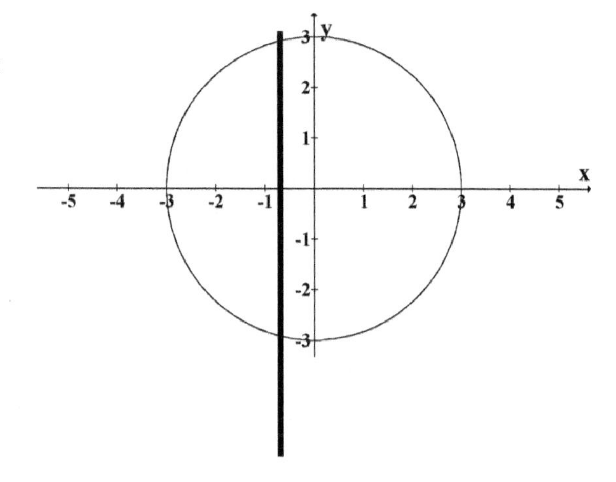

19. Sketch a graph that is a function, and one that is not a function. Then use the vertical line test to illustrate that the latter is not a function.

20. Other than examples found in this activity, describe an example from real life in which a function might be more useful than a relation that is not a function.

Notes

Notes

Functions 2: Characteristics of Functions

Model 1: Domain, Range, Independent and Dependent Variables

- The **domain** of a function f is the set of possible inputs. For a function defined by a formula, this is the set of all real numbers for x (inputs), that produce real valued outputs, $f(x)$.
- The **range** of the function is the set of all possible values of $f(x)$ given the domain.
- The input, x, is often called the **independent variable**.
- The output, $f(x)$, is often called the **dependent variable.**

Construct Your Understanding Questions (to do in class)

1. **Function A** accepts an input of <u>any English noun that can be made plural by adding an "S"</u>. (See examples at right.)

PEN → A → PENS

   a. Does the underlined phrase in the line above describe the **domain** or **range** of this function? [circle one]

   b. Does this underlined phrase describe the **independent** or **dependent** variable for this function? [circle one]

   c. Is the word "WISH" in the domain of Function A? Explain your reasoning.

2. Describe in words the domain of each function

   a. $f(x) = \frac{1}{x}$

   b. $f(x) = \sqrt{x-1}$

   c. $f(x) = \frac{1}{x-3}$

   d. $f(x) = x^2$

3. (Check your work) Only one of the functions in the previous question has a domain equal to all real numbers. Which one? For the other three, if you have not already done so, identify all real values of x that do <u>not</u> produce a real value of $f(x)$.

Model 2: Increasing or Decreasing Over an Interval

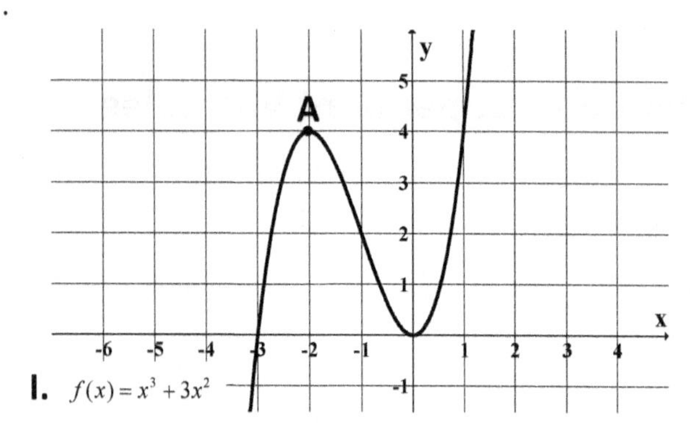

I. $f(x)=x^3+3x^2$

II. $f(x)=2x^3-3x^2-36x+12$

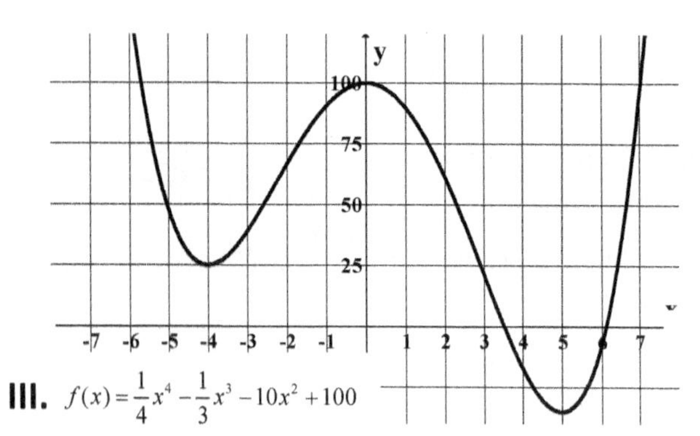

III. $f(x)=\frac{1}{4}x^4-\frac{1}{3}x^3-10x^2+100$

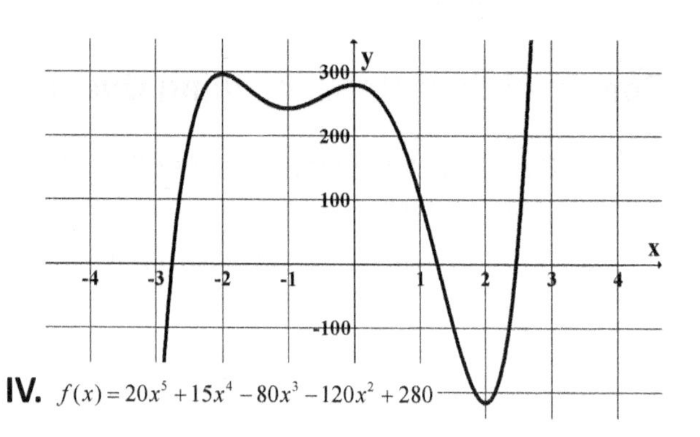

IV. $f(x)=20x^5+15x^4-80x^3-120x^2+280$

Construct Your Understanding Questions (to do in class)

4. Starting on the far left of Graph I, the **output values**, $f(x)$—plotted on the y axis, increase as you move from left to right until the point labeled **A** $(-2,4)$. They then decrease until the origin $(0,0)$, and increase again afterwards.

   a. Draw a dot (like the one labeled **A**) on each graph in Model 2 at each point where the function changes from increasing to decreasing or vice versa.

   b. For each graph, use the syntax below to write a statement identifying all the intervals on the x-axis where the function is **increasing** or **decreasing**. Graph I is done for you.

Example: Graph I. f is increasing for x in $(-\infty,-2)$, and $(0,\infty)$; decreasing for x in $(-2,0)$

5. The following phrases will complete the mathematical definitions in Summary Box F2.1. Write the correct ending in the correct box below.

    i. $f(x_1) > f(x_2)$ whenever $x_1 < x_2$ in I

    ii. $f(x_1) < f(x_2)$ whenever $x_1 < x_2$ in I

Summary Box F2.1: Increasing and Decreasing Functions

Description	Mathematical Definition
A function **is increasing** over **a given interval** if the values of $f(x)$ get larger as x gets larger.	The function f is **increasing** over an interval I if...
A function is **decreasing** over a given interval if the values of $f(x)$ get smaller as x gets larger.	The function f is **decreasing** over an interval I if...

Model 3: Concave Up and Concave Down

Later we will learn formal definitions of the terms **concave up (CU)** and **concave down (CD)**. For now, we will define these terms based the following examples.

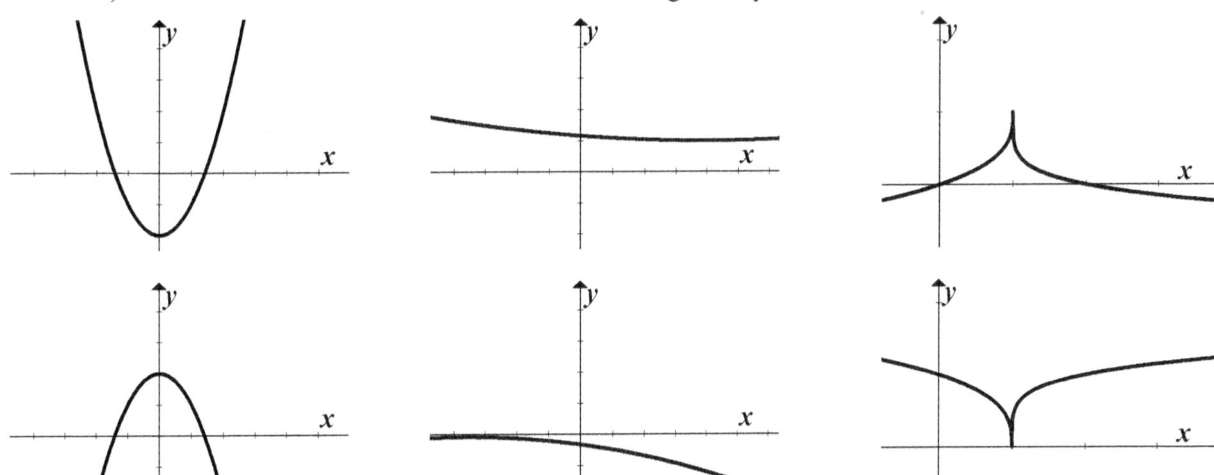

Construct Your Understanding Questions

6. The three graphs in the top row of Model 3 are **concave up (CU)**, and the three graphs in the bottom row are **concave down (CD)**. Describe in words...

    a. the curvature of the three graphs in the top row.

    b. the curvature of the three graphs in the bottom row.

7. Mark each point on the graph at right where concavity changes.

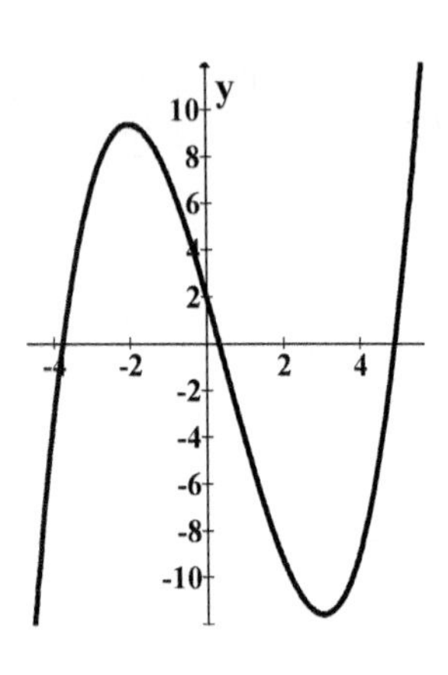

8. (Check your work) The point you were asked to mark in the previous question is called an **inflection point**. On the graph at right it is hard to tell *exactly* where the inflection point is, but you should have been able to tell that there is only one.

   The inflection point happens to be at $(0.5, 0)$.

9. Label each inflection point on the graphs from Model 2 (redrawn below).

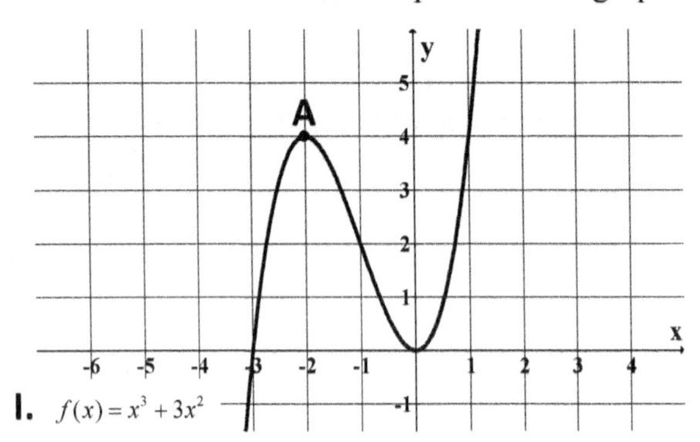

II. $f(x) = 2x^3 - 3x^2 - 36x + 12$

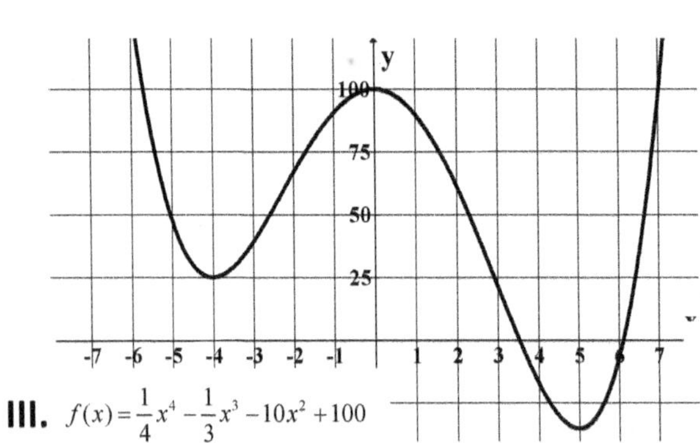

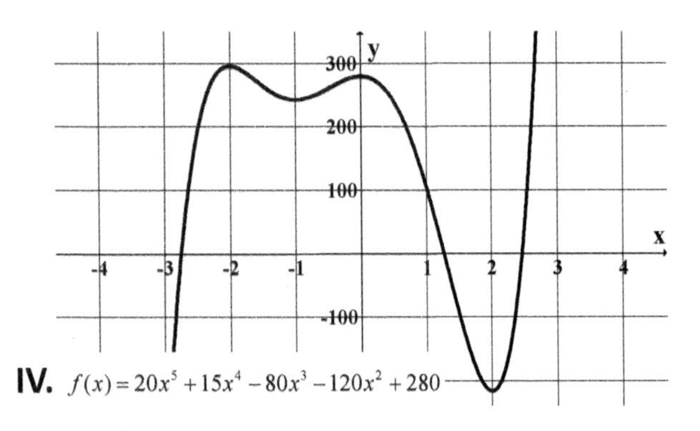

   a. (Check your work) There are a total of seven inflection points on these four graphs.

   b. Label each interval of concavity along the x-axis with **CU** or **CD**.

10. **True** or **False**: A graph that is concave up over an interval is also increasing over that same interval. If False, sketch the graph of a function that is decreasing and concave up over an interval (and label this interval d-CU).

11. On the axes below, plot a function $f(t)$ that describes the position of a car during the hypothetical situation described in a-g, below. The first two seconds are plotted for you. Take care to represent the function as increasing, decreasing, linear, concave up, or concave down.

    a. At $t = 0$, Cathy is driving at constant rate v_1 on a two lane road when she comes up on a slow moving tractor. There is no oncoming traffic and plenty of visibility, so at $t = 2$ seconds (without slowing down), she applies the accelerator and pulls into the opposing lane to pass.

    b. By $t = 4$ seconds she has doubled her rate to v_2. She stays at v_2 for 2 seconds while she passes the tractor.

    c. At $t = 6$ seconds she applies her brake and by $t = 8$ seconds she is traveling at a constant rate of v_1 again.

    d. She travels at v_1 for another 4 seconds when she hears a loud thump!

    e. Immediately (at $t = 12$ seconds), she applies the brake, and at $t = 14$ she comes to a stop on the shoulder.

    f. During the 4 seconds she is stopped, the tractor passes by with the farmer waving his arms and pointing back toward an object in the road.

    g. At $t = 18$ she accelerates in reverse for 1 second, then continues to back up slowly (at a constant velocity of v_3 for 3 more seconds when she slams on the brake, stopping almost immediately. (Continue your plot to $t = 24$ seconds.)

    To her surprise and ______________________, she sees that the object in the
    (write in an emotion here)

    road is ____.
    (complete the story as you see fit)

$y = f(t)$

Position of Cathy's Car

t (seconds)

2 4 6 8 10 12 14 16 18 20 22 24

12. Mark each section of your graph in the previous question that is increasing with "i", and decreasing with "d". Mark each section that is linear with "L", concave up with "CU", and concave down with "CD". For example, a section that is increasing and concave up would be marked "i-CU".

13. (Check your work) Identify each of the following sections on your graph in Question 11.
    a. Two sections that are "i-CD".
    b. One section that is "i-CU".
    c. One section that is "d-CD".
    d. Two sections that are linear and neither increasing nor decreasing.

14. There is a very small section of the graph in Question 11 that must be "d-CU". Identify this section and explain your reasoning.

Extend Your Understanding Questions (to do in or out of class)

15. For each statement, indicate whether a graph based on the statement (with the axis labels given) would be "i" or "d", and whether it will be "L", "CU" or "CD". Sketch a possible shape for the graph and briefly explain your reasoning. The first one is done for you.

Statement	x and y axis labels	Shape	Sketch of Graph (with reasoning)
The price of water in Honolulu is 0.218 cents per gallon. Plot a graph showing the amount of a monthly water bill as a function of the number of gallons used in a month.	x (gallons) y (cents or dollars)	i-L	y ($) x (gallons) The more water you use, the higher your monthly water bill, and the increase is linear because each gallon of water costs the same.
Renee is most efficient picking coffee beans at the start of the day when it is cool and she is feeling energetic. She picks slower and slower as the day goes on. Plot total pounds of beans picked by Renee as a function of time, in an 8 hour day.	x (hours) y (total beans)		

As your new president, I will immediately enact policies that slow depletion of the ozone layer. Plot ozone layer thickness as a function of time, going forward, assuming ozone depletion slows.	x (years) y (ozone layer thickness)		
A spherical chlorine tablet is dropped into a swimming pool. The amount of chlorine released is proportional to the tablet's surface area. Plot total chlorine released by this tablet as a function of time, as the tablet dissolves.	x (minutes) y (total chlorine released)		
A colony of bacteria with a large initial amount of food is a good model for unconstrained population growth (meaning that each bacterium produces many offspring). Plot the number of bacteria in a colony as a function of time, during the period of growth before food scarcity occurs.	x (hours) y (number of bacteria)		
Plot the number of bacteria in the previous example as a function of time, starting from the moment when food scarcity begins, through to the death of all individuals in the colony due to starvation. (Assume no food is ever provided after the large initial amount, as described above.)	x (hours) y (number of bacteria)	Indicate this on the graph since there is more than one shape represented. Be sure to clearly mark the beginning and end of each section that goes with a label.	

Notes

Functions 3: Compositions of Functions

Model 1: Word Machines

SIGN → A → SIGNS

COW → A → COWS

HI → A → HIS

BONK → B → KNOB

RAT → B → TAR

KAYAK → B → KAYAK

Construct Your Understanding Questions (to do in class)

1. Describe the effect of a) Machine A on a word b) Machine B on a word.

2. Consider a combination of Machines A and B such that the output of A is used as the input for B. See Example below, left. This is called a **composition** of these machines.

   a. Perform the same series of operations using "NIP" as the input (below, right).

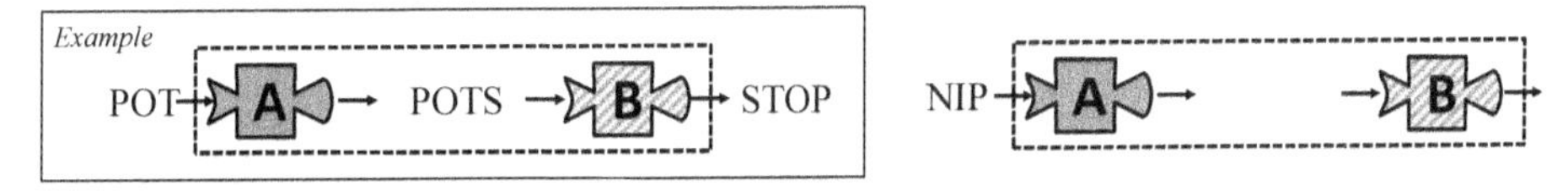

   b. By convention, the **composition** contained within the dotted box above is called "$B \circ A$" (which is read "B composed with A"). Label both dotted boxes above with the caption **Machine** $B \circ A$.

   c. For the composition in part a., which machine appears in the name first **A** or **B** [circle one]?

   d. Which machine is evaluated first, **A** or **B** [circle one]?

3. Using the naming convention from above, construct a name for this **composition** (shown two times below) and label each dotted box below with this name.

   a. Fill in the missing outputs of this machine.

   b. Does Machine $A \circ B$ (in this question) produce a different output from Machine $B \circ A$ (found in the previous question)? In other words, is the order of operation important?

Model 2: Composition of Functions

It is possible to **compose** functions as we composed machines on the previous page. The result, a composite function, can be written as shown below, or using the symbol $\circ$.

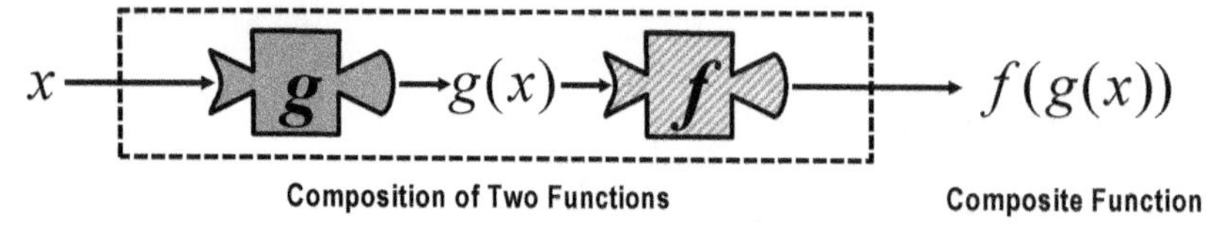

Composition of Two Functions — Composite Function

Construct Your Understanding Questions (to do in class)

4. Based on the naming conventions from the previous page, which is the correct name of the composite function in Model 2? $f \circ g$ or $g \circ f$ [circle one and explain your reasoning]

5. Assume $f(x) = x^2$ and $g(x) = x + 1$

   a. What is the output of $g \circ f$ with an input of $x = 3$?

      This can be written $(g \circ f)(3) =$

   b. What is the output of $f \circ g$ with an input of $x = 3$?

      This can be written $(f \circ g)(3) =$

   c. You can arrive at an algebraic expression for a composition by using x as the input. Fill in each of the following boxes to determine $(g \circ f)(x)$, and then $(f \circ g)(x)$. The first box is filled in for you.

      $x \rightarrow f \rightarrow$ [x^2] $\rightarrow g \rightarrow$ [ ]

      $(g \circ f)(x)$

      $x \rightarrow g \rightarrow$ [ ] $\rightarrow f \rightarrow$ [ ]

      $(f \circ g)(x)$

   d. For f and g as defined in this question, does $f \circ g = g \circ f$? Explain.

6. (Check your work) Substitute $x = 3$ into the equation you derived...
   a. for $(g \circ f)(x)$ and check that your answer is 10.
   b. for $(f \circ g)(x)$ and check that your answer is 16.

   Use this to also check your answers to parts a) and b) of Question 5. If any of your group's answers do not fit, check your methods against those of another group.

7. The function $f(x) = \sqrt{x+3}$ can be expressed as a composition of two simpler functions, *h* and *g* where $h(x) = \sqrt{x}$ and $g(x) = x + 3$.

   For these functions *f*, *g*, and *h*, does $f(x) = (g \circ h)(x)$ or $f(x) = (h \circ g)(x)$ [circle one]?

8. Consider the function $f(x) = (x-1)^2$. Propose two functions *g* and *h* such that $(g \circ h)(x) = f(x) = (x-1)^2$.

9. Later in this course we will discover that it is very useful to ***de*compose** a function into simpler functions (as you were asked to do in the previous two questions). For the following list of functions, propose two functions $g(x)$ and $h(x)$ such that $f(x) = (g \circ h)(x)$. You may not use the functions $f(x) = x$ or $g(x) = x$. (Construct an explanation for why using the function x would make this question too easy.)

   a. $f(x) = (x+4)^3$

   d. $f(x) = \sqrt{\frac{x}{2}}$

   b. $f(x) = \sqrt{x^2 + 7x}$

   e. $f(x) = x^3 + 4$

   c. $f(x) = \frac{\sqrt{x}}{2}$

   f. $f(x) = \frac{\sqrt{x-1}}{1+\sqrt{x-1}}$

Model 3: Shifting

At right is the parabola $h(x) = x^2$, along with four related parabolas, each one identical to $h(x) = x^2$, but shifted by a distance of 3 units up, down, right, or left, respectively.

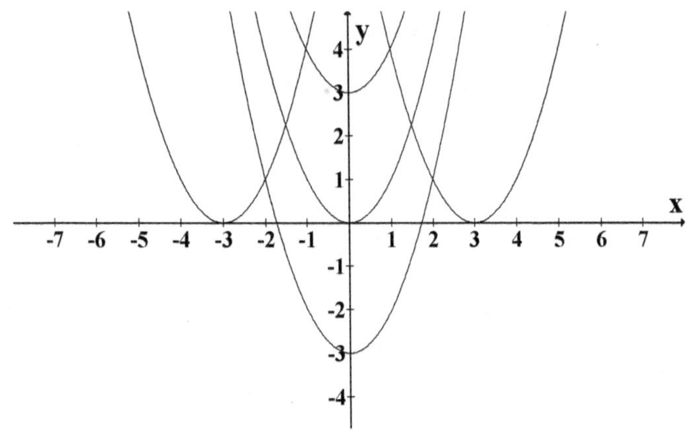

Construct Your Understanding Questions (to do in class)

10. In Model 3, label the graph of $h(x) = x^2$ and the graphs of the four other functions listed in the table below. Check your work by finding the vertex of each parabola and confirming that this point is satisfied by the equation you chose as a match.

    a. Complete Column 2 by filling in the word LEFT, RIGHT, UP, or DOWN.

    b. Complete Column 3: Each shift of $h(x) = x^2$ can be represented as a composition of the original function $h(x) = x^2$ and another function $g(x)$. For each row, propose a function $g(x)$ that accomplishes this shift. [The first one is done for you.]

    c. In Column 4 indicate if $h \circ g$ or $g \circ h$ give the function in Column 1.

Function	Direction of Shift from $h(x) = x^2$	$g(x)$ that can accomplish this shift	$h \circ g$ or $g \circ h$ [indicate which] gives $f(x)$ shown in Column 1
$f(x) = x^2 + 3$		$g(x) = x + 3$	
$f(x) = x^2 - 3$			
$f(x) = (x + 3)^2$			
$f(x) = (x - 3)^2$			

11. (Check your work) $f(x)$ shown at right is the graph of a sine wave shifted up by 3 units.

    Assume $h(x) = \sin x$.

    Which function $g(x)$ from the previous question and which composition $h \circ g$ or $g \circ h$ or **both** [circle all that apply] will give the function $f(x)$ shown at right?

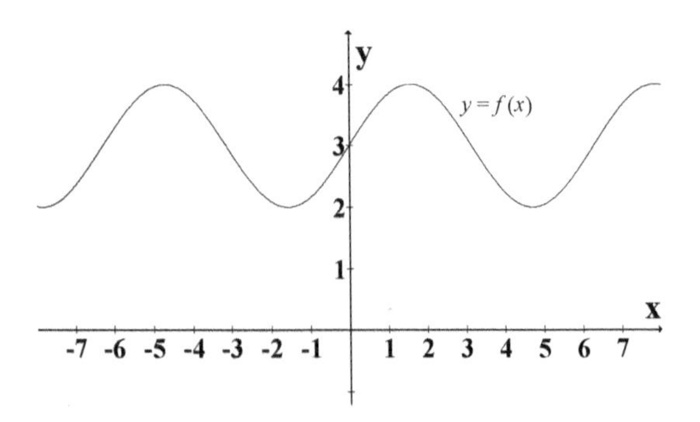

12. (Check your work) In general, if you want to shift a function $h(x)$ a vertical distance c (down or up) to produce a new shifted function $f(x)$, you can compose $h(x)$ with another function $g(x) = x + c$ to give $f(x)$.

    a. (Check your work) Check that this information is consistent with your answer to the previous question.

    b. c in the function $g(x) = x + c$ can be positive or negative. Describe the result of the composition when c is positive versus negative.

    c. Which composition $h \circ g$ or $g \circ h$ or **both** [circle all that apply] will give the function $f(x)$, as described above?

13. Suppose you want to shift a function $h(x)$ a horizontal distance c (left or right) to produce a new shifted function $f(x)$. Write a function $g(x)$ and an expression stating how $h(x)$ and $g(x)$ can be composed to give $f(x)$. [Hint: look back at the horizontal shifts in Question 10.]

14. Assume: $h(x) = x^2$ $g_v(x) = x + 1$ and $g_h(x) = x - 2$

    a. Describe the effect of the composition $g_v \circ h \circ g_h$ in terms of a) horizontal and b) vertical shifting of the graph of the parabola in comparison to the original function h.

    b. (Check your work) Is your answer to part a) consistent with the fact that this shifted parabola will have its vertex at the point (2,1)?

    c. Evaluate the composition $(g_v \circ h \circ g_h)(x)$ so as to generate the equation in terms of x for this shifted parabola. Remember to evaluate the functions in the name from right to left (i.e., evaluate g_h first).

15. (Check your work) Is your answer to the previous question consistent with Summary Box F3.1?

Summary Box F3.1: Shifting as Described by Compositions of Functions

When $g_h(x) = x + c_h$ and $g_v(x) = x + c_v$ the composition $g_v \circ h \circ g_h$ shifts the function $h(x)$...

- in the horizontal direction by an amount c_h
- in the vertical direction by an amount c_v

16. In Summary Box F3.1...

 a. when c_h is positive, the horizontal shift is **right** or **left** [circle one].

 b. when c_v is positive, the vertical shift is **up** or **down** [circle one].

Model 4: Reflecting About the x and y Axes

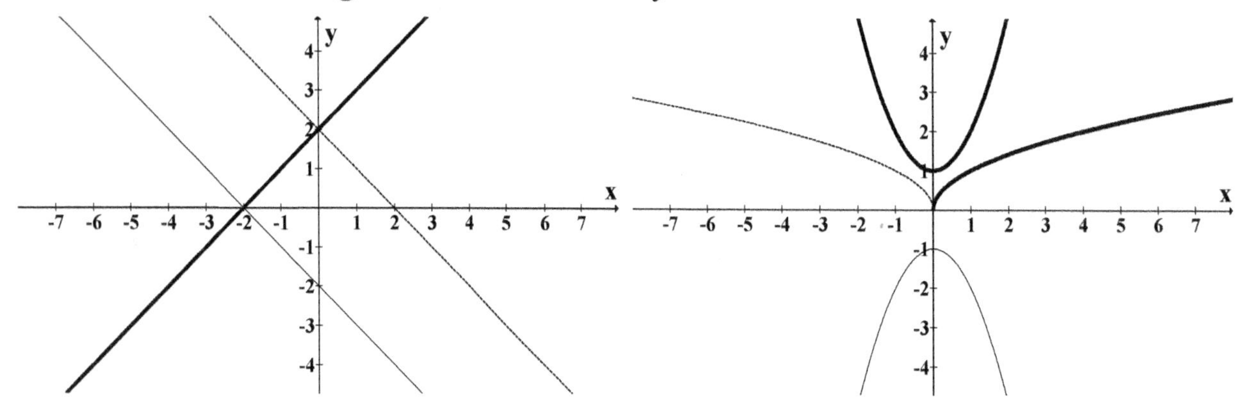

Extend Your Understanding Questions (to do in or out of class)

17. Identify the graph of the function $f(x) = x + 2$ in Model 4.

 a. Which line is a reflection of $f(x)$ about the <u>x axis:</u>
 The **light solid line** or the **dashed line** [circle one] in Model 4?

 b. Which line is a reflection of $f(x)$ about the <u>y axis:</u>
 The **light solid line** or the **dashed line** [circle one] in Model 4?

 c. Write an equation for each of these lines.

18. (Check your work) Are your answers on the previous page consistent with the following?

    The functions $f(x)=x+2$ and $-f(x)=-x-2$ are reflections of each other about the x axis.

    The functions $f(x)=x+2$ and $f(-x)=-x+2$ are reflections of each other about the y axis.

19. The functions $f(x)=x^2+1$ and $f(x)=\sqrt{x}$ are shown in **bold** on the right of Model 4.
    a. Identify the equations of the light solid curve, and the dotted curve on this graph.

    b. The reflection of $f(x)=\sqrt{x}$ about the x axis is not shown in Model 4. Add it and label it with its equation.

20. A function $f(x)$ can be composed with the function $g(x)=-x$ to give a reflection of $f(x)$.
    a. Which composition ($f \circ g$ <u>or</u> $g \circ f$) gives the reflection of $f(x)$ about the x axis? [circle one]

    b. Which composition ($f \circ g$ <u>or</u> $g \circ f$) gives the reflection of $f(x)$ about the y axis? [circle one]

21. (Check your work) Check your conclusions above with at least one other group, then complete Summary Box F3.2 by filling in each blank with $g \circ f$ or $f \circ g$, as appropriate.

Summary Box F3.2: Reflecting of Functions

For a function $f(x)$, a composition with the function $g(x)=-x$ results in a reflection about the...

- x axis when the composition is __________________

- y axis when the composition is __________________

Model 5: Stretching and Compressing of Functions

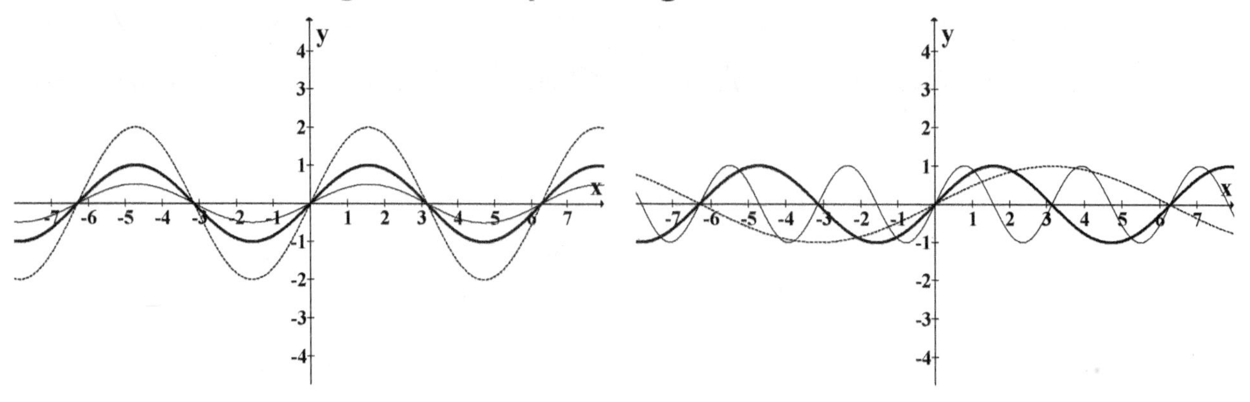

Extend Your Understanding Questions (to do in or out of class)

22. Each graph in Model 5 shows the function $h(x) = \sin x$ in **bold**. Determine which graph in Model 5 shows $h(x)$ along with versions of $h(x)$ that are stretched and compressed in ....

    a. the vertical direction. (label this graph "vertical stretch/compress")

    b. the horizontal direction. (label this graph "horizontal stretch/compress")

23. Complete the table in Summary Box F3.3 by writing the correct pair of terms in each empty box. Choose from: **compressed vertically**, **compressed horizontally**, **stretched vertically**, **stretched horizontally**.

Summary Box F3.3: Stretching and Compressing of Functions

For a function $h(x)$, a composition with the function $g(x) = cx$ (where c is positive) results in stretching or compressing of the original function to give a new function $f(x)$.

	$0 < c < 1$ (e.g. ½)	$c > 1$ (e.g. 2)
$f = g \circ h$	Compared to $h(x)$, $f(x)$ is...	Compared to $h(x)$, $f(x)$ is...
$f = h \circ g$	Compared to $h(x)$, $f(x)$ is...	Compared to $h(x)$, $f(x)$ is...

Confirm Your Understanding Questions (to do at home)

24. Use a graphing program to plot the graph of the function $f(x) = ax^2$ where $a = 1$. Then also plot the graph of the function $f(x) = ax^2$ where the value of a is given below. You will have 4 graphs when you are finished. For each value of a, describe in words the effect of changing a from $a = 1$ to...
    a. $a = -1$
    b. $a = 3$
    c. $a = 10$

25. Use a graphing program to plot the function $f(x) = (x + c)^2$ where $c = 0$ and the changes noted below. For each change, describe in words the effect of changing c from $c = 0$ to...
    a. $c = -1$
    b. $c = 3$
    c. $c = 10$

26. Use a graphing program to plot the function $f(x) = x^2 + d$ where $d = 0$ and the changes noted below. For each change, describe in words the effect of changing d from $d = 0$ to...
    a. $d = -1$
    b. $d = 3$
    c. $d = 10$

27. Write the equation the parabola $f(x) = x^2$ (shown below) shifted a distance c in the direction indicated by each arrow.

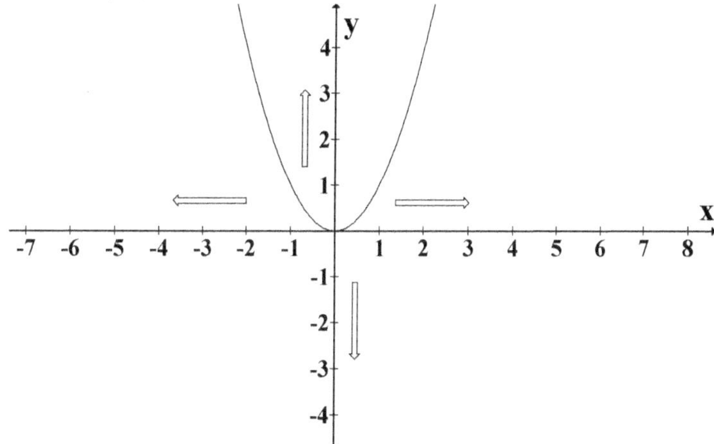

Notes

Limits 1: Limit of a Function

Model 1: Guessing a Limit of a Function from a Graph or Table

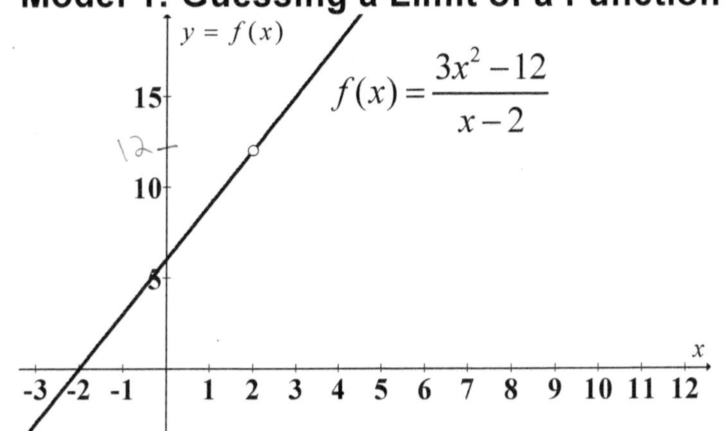

x	$y = f(x)$
1	7
1.9	11.7
1.99	11.97
1.999	11.997
2	undefined
2.001	12.003
2.01	12.03
2.1	12.3
3	14

Construct Your Understanding Questions (to do in class)

1. For the function in Model 1, what is... (add your answers to the empty boxes on the table)

   a. $f(1) =$ b. $f(2) =$ c. $f(3) =$

2. For $f(x)$ in Model 1...

   a. What is the meaning of the open circle on the graph at the point where $x = 2$? (Check your work) Does this confirm your answer to Question 1b? Explain.

   undefined

   b. Use the table to guess what y value $f(x)$ is approaching as x approaches 2. We call this the "**limit** of $f(x)$ as x approaches 2", which is written... $\lim_{x \to 2} f(x)$

   $\lim_{x \to 2} f(x) = 12$

   c. The y value associated with the open circle on the graph in Model 1 is 12. Make a hash mark for this value of y at the appropriate place on y-axis in Model 1, and label it with a 12. (Check your work) Is this consistent with your answer to Question 2b?

3. Guess each limit using the graph.

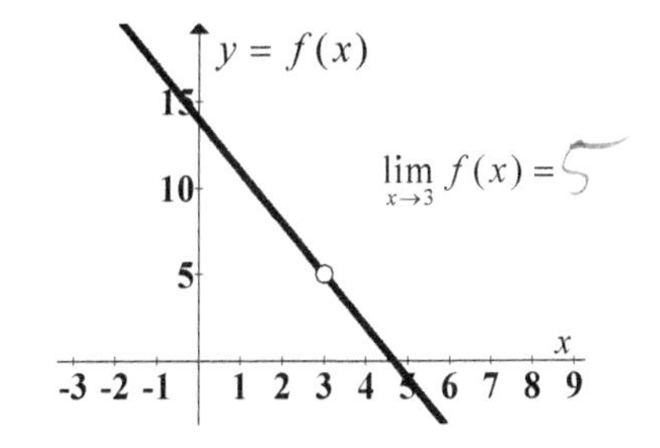

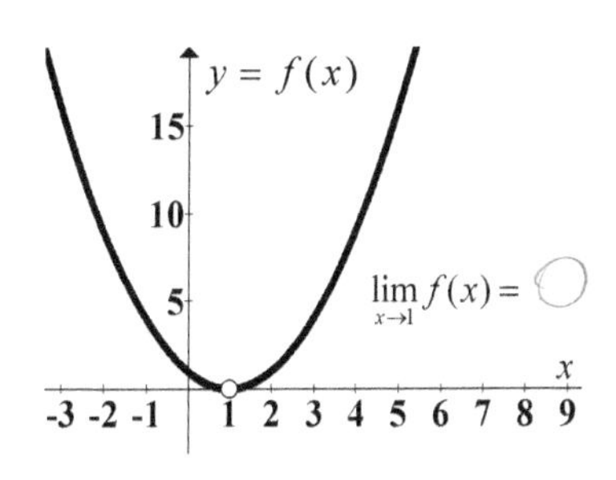

4. Most questions in this activity focus on limits at a point where the y value of the function is *not* defined. However, in this question we ask you to find limits at points where the y value *is* defined. These are often easy to find because, for these functions, $\lim_{x \to a} f(x) = f(a)$.

   a. For the function in **Model 1**, what is:

   $\lim_{x \to 3} f(x) =$ 14 $\qquad \lim_{x \to 1} f(x) =$ 7 $\qquad \lim_{x \to 0} f(x) =$ 6

   b. Based on the graph at right, what is …

   $\lim_{x \to -1} g(x) =$ 17

   $\lim_{x \to 0} g(x) =$ 14

   $\lim_{x \to 5} g(x) =$ −1

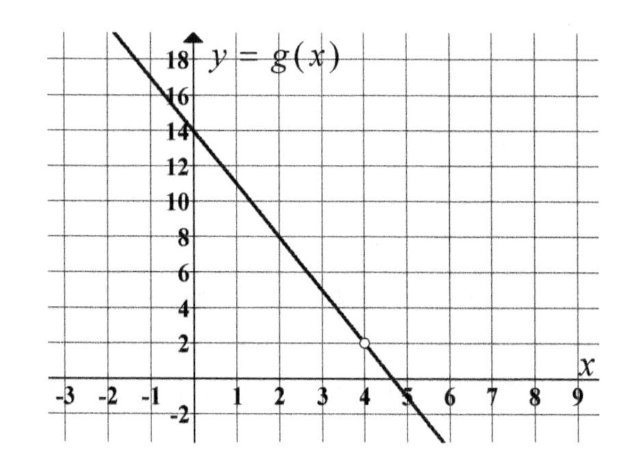

5. Fill in the blank to make the statement true:

   For the function shown on the graph in part b of the previous question, $\lim_{x \to a} g(x) = g(a)$ for every value of a except $a =$ 4

6. Exactly <u>one</u> of the following statements is FALSE. Cross out the false statement and match each true statement to the graph (A or B) that demonstrates that it is true. Hint: First look at the Graphs A and B, and find values of $x = a$ where $f(a)$ does not exist or $\lim_{x \to a} f(x) \neq f(a)$.

   i. The $\lim_{x \to a} f(x)$ can exist even if $f(a)$ does not exist.

   ii. The $\lim_{x \to a} f(x)$ does not exist if $f(a)$ exists.

   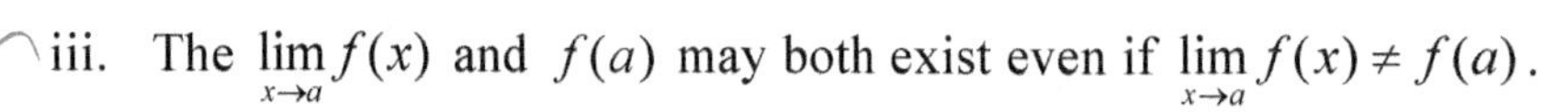

   iii. The $\lim_{x \to a} f(x)$ and $f(a)$ may both exist even if $\lim_{x \to a} f(x) \neq f(a)$.

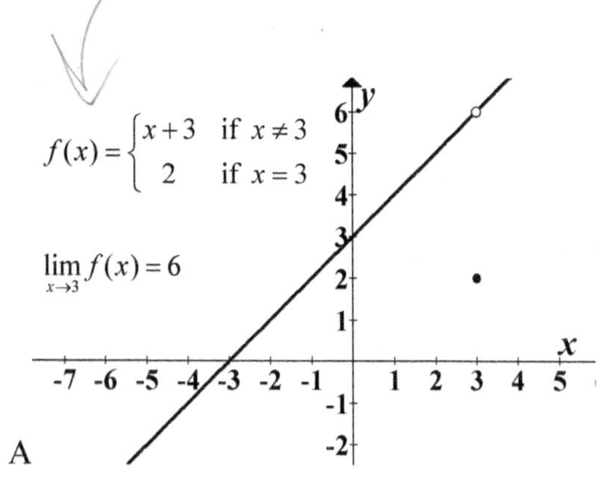

A

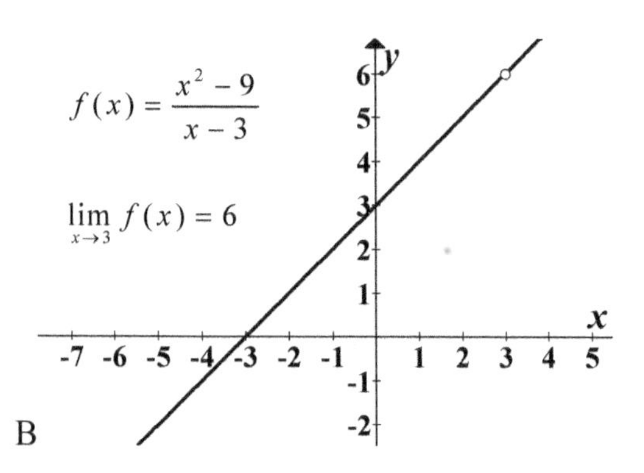

B

7. Guess $f(a)$ and $\lim_{x\to a} f(x)$ requested on each graph. (Dotted lines mark vertical asymptotes.)

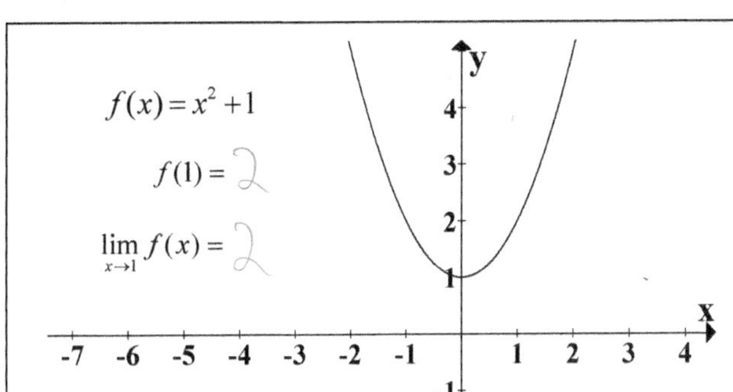

$f(x)=\frac{x^3-x^2+x-1}{x-1}$

$f(1)=$

$\lim_{x\to 1} f(x)=$

b.

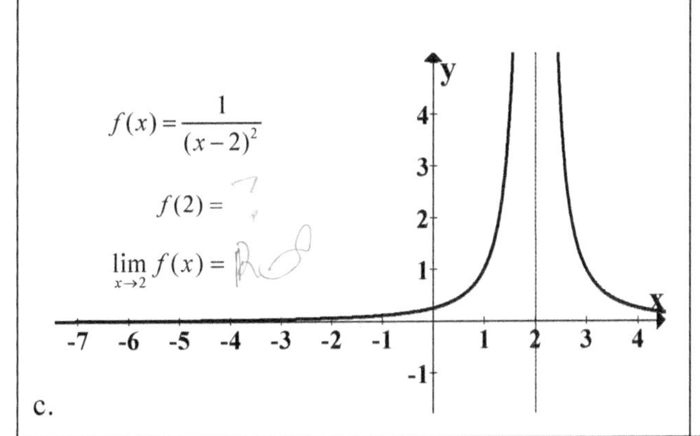

$f(x)=\ln(x-1)$

$f(1)=$

$\lim_{x\to 1} f(x)=$

d.

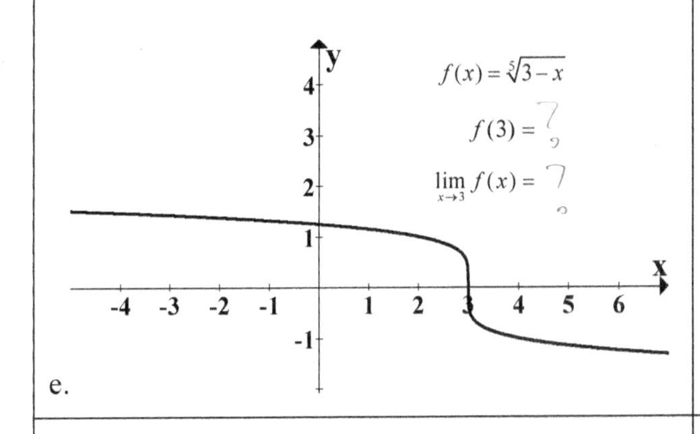

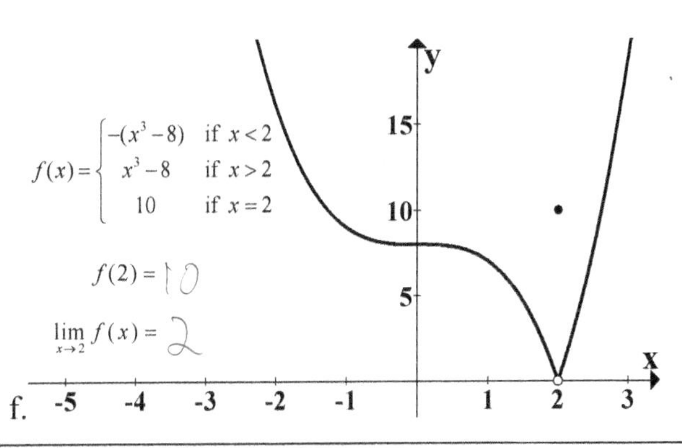

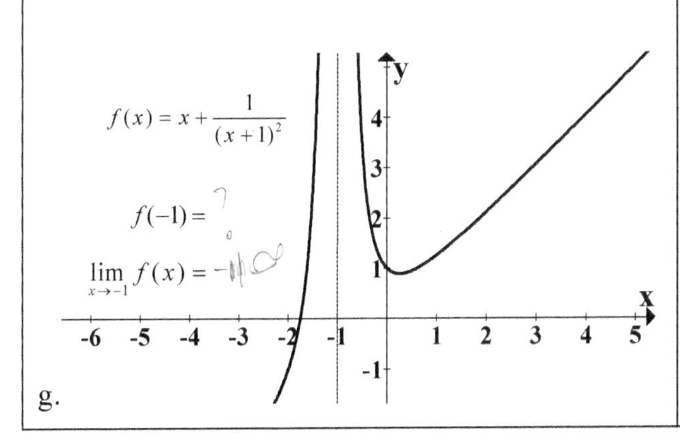

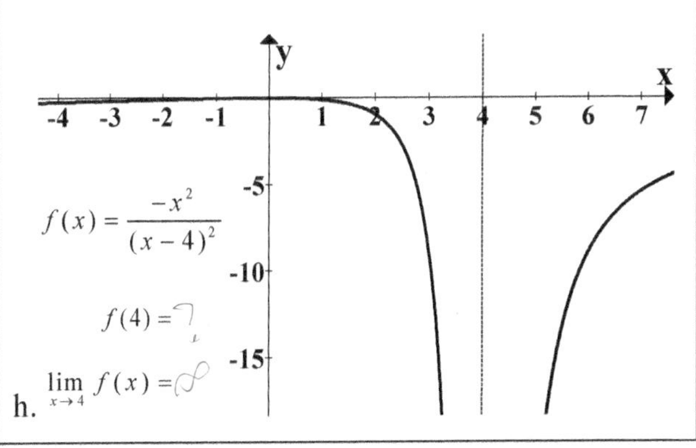

8. (Check your work) Four of the graphs in the previous question ask for a limit that is not a real number.

   a. Identify these four graphs and explain your reasoning.

   b. Is your answer to part a of this question consistent with Summary Box L1.1? If not, write down any questions you have and address these with another group or the instructor.

Summary Box L1.1: Infinite Limits

Infinity is not a number, so we say that if $\lim_{x \to a} f(x) = \infty$ (or $-\infty$) then the limit does not exist.

To many students, this language seems contradictory. Students want to know: "Is the limit ∞ or does it not exist?" The answer is: both.

Think of it this way: Within the broader category of limits that do not exist, those that equal infinity have something in common with one another, as do those that equal minus infinity, and this helps us describe the graph. Each therefore makes up a sub-category within the set of limits that do not exist.

9. Consider the graph at right.

$$f(x) = \begin{cases} 1 & \text{if } x < 2 \\ 3 & \text{if } x \geq 2 \end{cases}$$

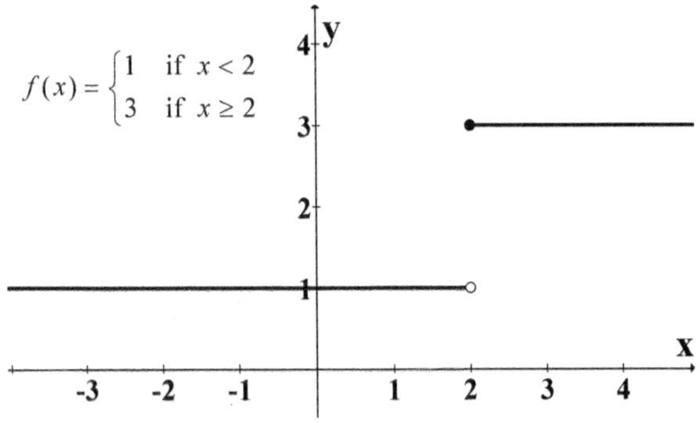

   a. Guess the limit of $f(x)$ as x approaches 2 from the left (e.g. as x goes from 1.9 to 1.99 to 1.999 etc.).

   Explain your reasoning.

   b. Guess the limit of $f(x)$ as x approaches 2 from the right.

   c. Identify the symbol in Summary Box L1.2 (on the next page) used to specify that a limit is a left-hand limit.

Summary Box L1.2: Left-Hand Limit

$\lim_{x \to a^-} f(x) =$ the limit of $f(x)$ as x approaches a from the left, that is when $x < a$.

This is also called the **left-hand limit** of $f(x)$ as x approaches a.

(Informal) Definition of a Left Hand Limit

$\lim_{x \to a^-} f(x) = L$ if all values of $f(x)$ are within any given distance of L for all values of x within a corresponding small distance of a, but always less than a.

10. Based on the notation above, guess the notation for the right-hand limit of a function, and use this notation to report the limit of the function in Question 9 as x approaches 2 from the right (e.g. as x goes from 2.1 to 2.01 to 2.001 etc.).

$\lim_{x \to 2^+} f(x) = 3$

11. Guess the left-hand limit and right-hand limit as $x \to 1$ for each function. Write these below each graph using proper notation. (Note: This question is continued on the next page.)

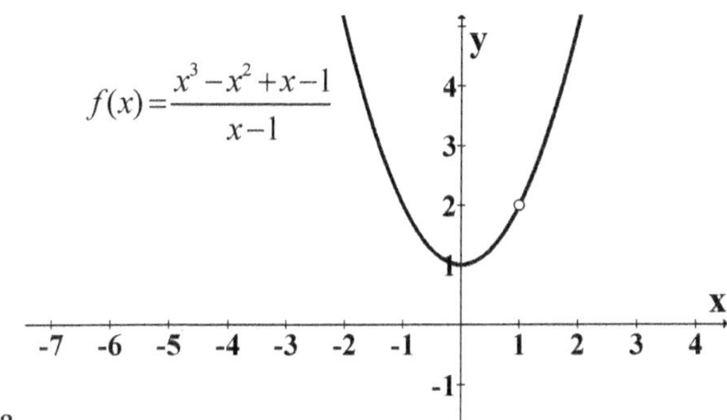

a.

$\lim_{x \to 1^-} f(x) = 2$

$\lim_{x \to 1^+} f(x) = 2$

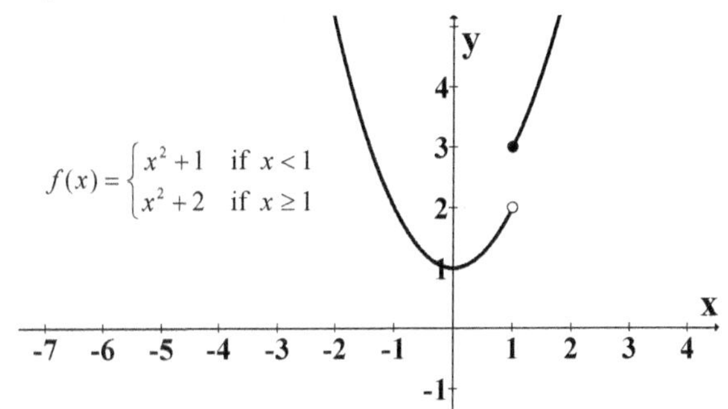

b.

$\lim_{x \to 1^-} f(x) = 2$

$\lim_{x \to 1^+} f(x) = 3$

$f(x) = \begin{cases} x^2 + 1 & \text{if } x < 1 \\ -x + 3 & \text{if } x \geq 1 \end{cases}$

c.

$\lim_{x \to 1^-} f(x) = 2$

$\lim_{x \to 1^+} f(x) = 2$

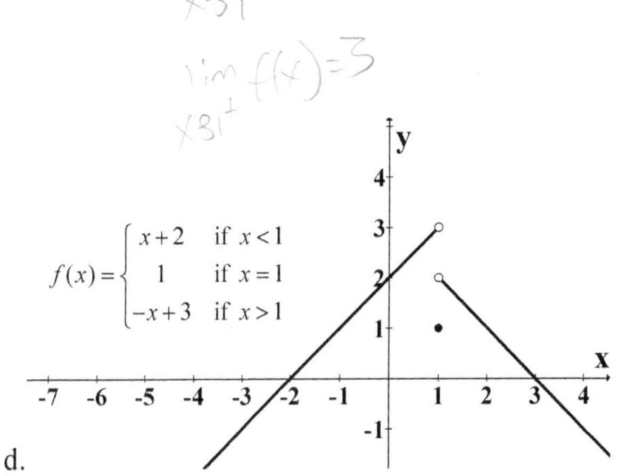

d.

$\lim_{x \to 1^-} f(x) = 3$

$\lim_{x \to 1^+} f(x) = 2$

e.

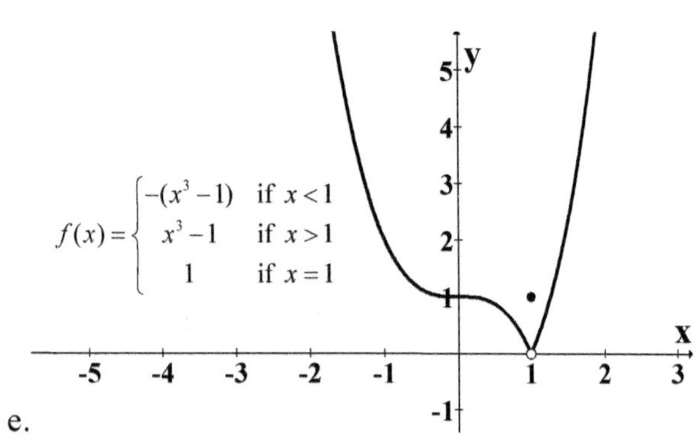

f.

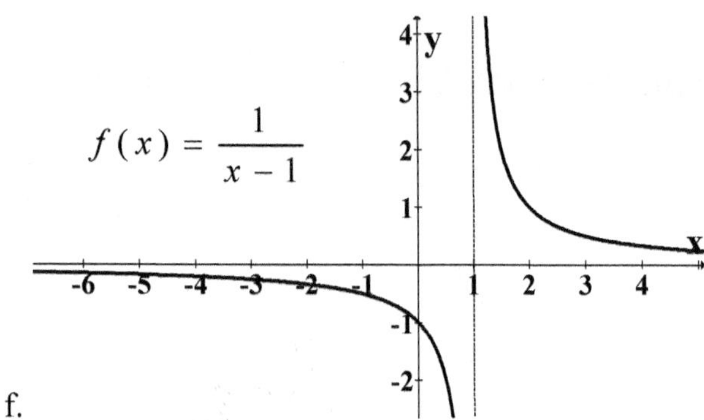

g.

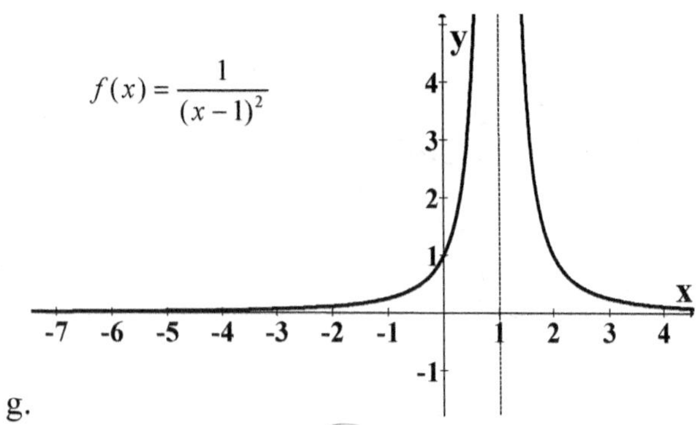

h.

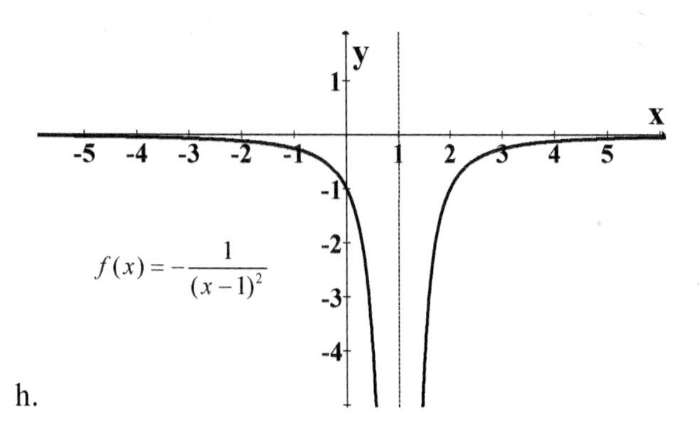

12. Identify each graph in the previous question for which $\lim_{x\to 1} f(x)$ does not exist, then label each of these with a reason. Choose from "$\lim_{x\to a^-} f(x) \neq \lim_{x\to a^+} f(x)$" or "$L$ not a real number."

Summary Box L1.3: Two-Sided Limit

$\lim_{x\to a} f(x) = L$ exists if and only if $\lim_{x\to a^-} f(x) = \lim_{x\to a^+} f(x) = L$, where L is a real number.

13. (Check your work) A student says the $\lim_{x\to 1} f(x)$ for Graphs g and h of Question 11 exists because $\lim_{x\to 1^-} f(x) = \lim_{x\to 1^+} f(x)$. Explain both the logic and the error(s) in this student's claim.

14. Based on the table shown at right, guess

$$\lim_{x\to 0}\left(\sin\frac{\pi}{x}\right)=$$

Explain your reasoning.

x	$\sin\frac{\pi}{x}$
2	1
1	0
0.1	0
0.01	0
0.001	0
-0.001	0
-0.01	0
-0.1	0
-1	0
-2	-1

15. Two graphs of $f(x)=\sin\frac{\pi}{x}$ are provided, below.

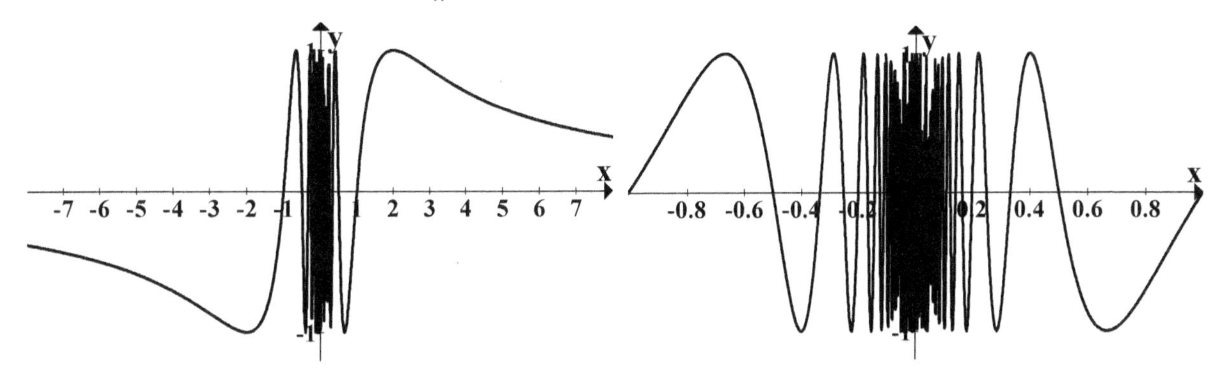

a. Describe how the x axis differs between these two graphs.

b. Use the graphs to estimate the value of f when x is…

i. 2   ii. $\frac{2}{3}\approx 0.67$ iii. $\frac{2}{5}=0.4$ iv. -2 v. $-\frac{2}{3}\approx -0.67$ vi. $-\frac{2}{5}=-0.4$

c. Based on the pattern in part b, guess the value of the function when x is…

i. $\frac{2}{7}$ ii. $-\frac{2}{7}$

d. (Check your work) Are your guesses in part c consistent with the graphs?

Note that $\frac{2}{7}\approx 0.29$

e. Based on these graphs, describe what appears to be happening to $y=f(x)$ as x approaches zero.

16. Complete the last column on the table at right. Do the entries in this column confirm your answer to part e of the previous question? That is, according to all the information you have about this function, what appears to be happening to $f(x)$ as x approaches zero?

x *(decimal)*	x *(fraction)*	$\frac{\pi}{x}$	$\sin\frac{\pi}{x}$
2.0000	2	$\frac{\pi}{2}$	
0.6667	$\frac{2}{3}$	$3\cdot\frac{\pi}{2}$	
0.4000	$\frac{2}{5}$	$5\cdot\frac{\pi}{2}$	
0.2857	$\frac{2}{7}$	$7\cdot\frac{\pi}{2}$	
0.2222	$\frac{2}{9}$	$9\cdot\frac{\pi}{2}$	1
0.1818	$\frac{2}{11}$	$11\cdot\frac{\pi}{2}$	-1
0.1538	$\frac{2}{13}$	$13\cdot\frac{\pi}{2}$	1
0.0571	$\frac{2}{35}$	$35\cdot\frac{\pi}{2}$	-1
0.0444	$\frac{2}{45}$	$45\cdot\frac{\pi}{2}$	1
0	0	*undefined*	*undefined*
-0.0444	$-\frac{2}{45}$	$-45\cdot\frac{\pi}{2}$	-1
-0.0571	$-\frac{2}{35}$	$-35\cdot\frac{\pi}{2}$	1
-0.1538	$-\frac{2}{13}$	$-13\cdot\frac{\pi}{2}$	-1
-0.1818	$-\frac{2}{11}$	$-11\cdot\frac{\pi}{2}$	1
-0.2222	$-\frac{2}{9}$	$-9\cdot\frac{\pi}{2}$	-1
-0.2857	$-\frac{2}{7}$	$-7\cdot\frac{\pi}{2}$	1
-0.4000	$-\frac{2}{5}$	$-5\cdot\frac{\pi}{2}$	-1
-0.6667	$-\frac{2}{3}$	$-3\cdot\frac{\pi}{2}$	1
-2.0000	-2	$-\frac{\pi}{2}$	-1

17. Construct an explanation for how the table at the top of the previous page could mislead someone to conclude that $\lim_{x\to 0}\left(\sin\frac{\pi}{x}\right)=0$.

18. (Check your work) Construct an explanation for why $\lim_{x\to 0}\left(\sin\frac{\pi}{x}\right)$ does not exist.

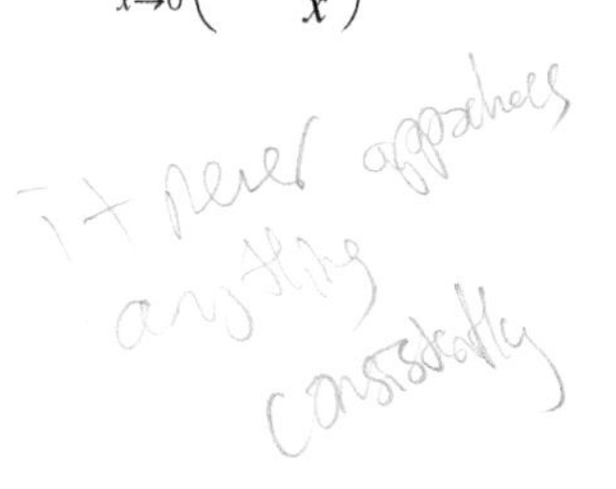

19. Review your answers to Question 12 and briefly describe how $\lim_{x\to 0}\left(\sin\frac{\pi}{x}\right)$ is different from all other limits found in this activity that do not exist.

Summary Box L1.4: One Informal Definition of a Limit

For a function f that is defined on both sides of a, but not necessarily at a, the $\lim_{x \to a} f(x) = L$ if all values of $f(x)$ are within any given distance of L whenever x is within a corresponding small distance of a, but not equal to a.

(In a later activity a more precise definition of a limit will be developed.)

Notes

Limits 2: Limit Laws

Model 1: Addition of Functions

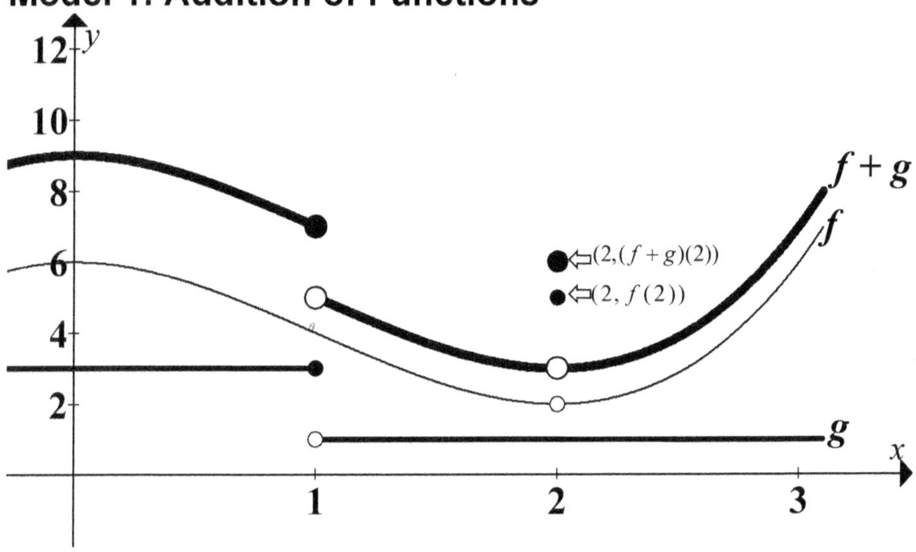

The thin curve represents the function f . The bold step function is g, and the extra-bold function is $(f+g)(x)$, where $(f+g)(x)=f(x)+g(x)$.

Construct Your Understanding Questions (to do in class)

1. According to the graph of f what is

   a. $f(1)=$ b. $f(2)=$

2. Check your answers to Question 1. Then use Model 1 to fill in missing entries.

a	$f(a)$	$\lim_{x\to a} f(x)$	$g(a)$	$\lim_{x\to a} g(x)$	$(f+g)(a)$	
0	6	6	3	3	9	
1	4	4	3			
2	5					
3	6		1		7	

3. (Check your work) One cell in the table above is a limit that is undefined. Circle this cell.

4. Add the heading $\lim_{x\to a}(f+g)(x)$ to the top of the shaded column, and fill in this column for $a = 0$, 1, 2, and 3.

5. Which two columns are related to the shaded column? Describe how they are related.

6. (Check your work) Which Limit Law in Summary Box L2.1 confirms your answer to the previous question? (If none fit, consider revising your answer to the previous question.)

Summary Box L2.1: Limit Laws

If $\lim_{x \to a} f(x)$ and $\lim_{x \to a} g(x)$ exist then…

1. $\lim_{x \to a}[f(x)+g(x)] = \lim_{x \to a} f(x) + \lim_{x \to a} g(x)$ and $\lim_{x \to a}[f(x)-g(x)] = \lim_{x \to a} f(x) - \lim_{x \to a} g(x)$
2. $\lim_{x \to a}[c \cdot f(x)] = c \cdot \lim_{x \to a} f(x)$ where c is a constant
3. $\lim_{x \to a}[f(x) \cdot g(x)] = \lim_{x \to a} f(x) \cdot \lim_{x \to a} g(x)$
4. $\lim_{x \to a}\left[\dfrac{f(x)}{g(x)}\right] = \dfrac{\lim_{x \to a} f(x)}{\lim_{x \to a} g(x)}$ provided $\lim_{x \to a} g(x) \neq 0$
5. $\lim_{x \to a} x = a$
6. $\lim_{x \to a} c = c$ where c is a constant
7. $\lim_{x \to a}[f(x)]^n = \left[\lim_{x \to a} f(x)\right]^n$ where n is a positive integer
8. $\lim_{x \to a} \sqrt[n]{f(x)} = \sqrt[n]{\lim_{x \to a} f(x)}$ n is a positive integer; $\lim_{x \to a} f(x) > 0$ when n is even

7. What is… (For each, give the number of the Limit Law that confirms your answer.)

a. $\lim_{x \to 2} x =$ b. $\lim_{x \to 42} x =$ c. $\lim_{x \to 0} 12 =$ d. $\lim_{x \to 1000} 12 =$

8. (Check your work) The two functions in the previous question are shown on the graph at right as f and g.

a. Give the formulas: $f(x) =$ $g(x) =$

b. Are your answers to Question 7 consistent with these graphs? If not, go back and check your work.

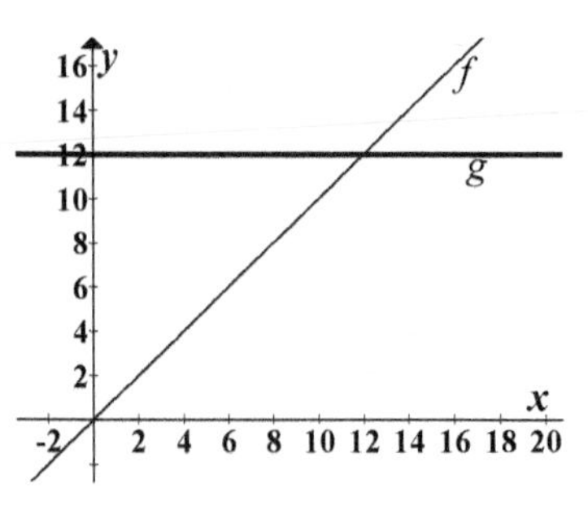

9. Let $\lim_{x\to 0} p(x) = 2$ and $\lim_{x\to 0} q(x) = 10$.

Use the Limit Laws in Summary Box L2.1 to find each of the following. In each box, <u>write the number of each Limit Law that you used</u>. The first one is done for you.

a. $\lim_{x\to 0}[q(x) - p(x)] = 8$ ***Limit Law 1***	b. $\lim_{x\to 0}[q(x)\cdot p(x)] =$ 20 law 3
c. $\lim_{x\to 0}[7\cdot p(x)] =$ 14 law 2	d. $\lim_{x\to 0}\left[\frac{q(x)}{p(x)}\right] =$ 5 law 4
e. $\lim_{x\to 0}[q(x)]^3 =$ 1000 1000 law 7	f. $\lim_{x\to 0}\left[\sqrt[3]{50\cdot q(x)\cdot p(x)}\right] =$ 10 law 8

10. Consider the limit and work shown in the large box at right. For each line, give the Limit Law that was used to generate the expression from the one above it. One is done for you.

$$\lim_{x\to -1}\left[\frac{2x^2-5x+7}{x^2-6x}\right] =$$

$$= \frac{\lim_{x\to -1}(2x^2-5x+7)}{\lim_{x\to -1}(x^2-6x)}$$ law 4

$$= \frac{\lim_{x\to -1} 2x^2 - \lim_{x\to -1} 5x + \lim_{x\to -1} 7}{\lim_{x\to -1} x^2 - \lim_{x\to -1} 6x}$$ *Limit Law 1*

$$= \frac{2\lim_{x\to -1} x^2 - 5\lim_{x\to -1} x + \lim_{x\to -1} 7}{\lim_{x\to -1} x^2 - 6\lim_{x\to -1} x}$$ law 2

$$= \frac{2\left[\lim_{x\to -1} x\right]^2 - 5\lim_{x\to -1} x + \lim_{x\to -1} 7}{\left[\lim_{x\to -1} x\right]^2 - 6\lim_{x\to -1} x}$$ law 7

$$= \frac{2(-1)^2 - 5(-1) + \lim_{x\to -1} 7}{(-1)^2 - 6(-1)}$$

$$= \frac{2+5+7}{1+6} = 2$$

11. In the previous question, the value of the limit could have been found by simply substituting -1 for x in the original equation.

a. Construct an explanation for why this (substituting -2 for x) does not work for $\lim_{x\to -2}\frac{x^2-4}{x+2}$.

$-2+2$ is 0

b. (Check your work) Circle the part of Limit Law 4 that confirms your answer to part a of this question and explain your reasoning.

undefined at 0

12. Consider the limit in Question 10: $\lim_{x \to a} \frac{2x^2 - 5x + 7}{x^2 - 6x}$. Find the value(s) of a for which you cannot apply Limit Law 4. Show your work and explain your reasoning.

13. (Check your work) To answer the previous question you likely found the values of x for which $\frac{2x^2 - 5x + 7}{x^2 - 6x}$ is not defined. If you did not do this, discuss with your group why this strategy would lead you to the correct answer.

✓

14. The graph below shows the function f from Question 11, and a function g.

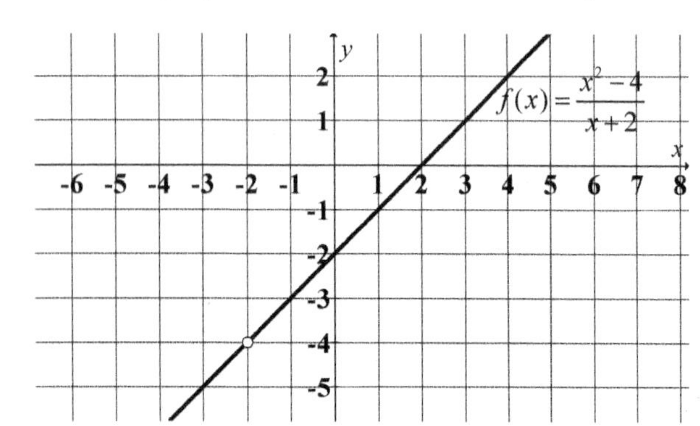

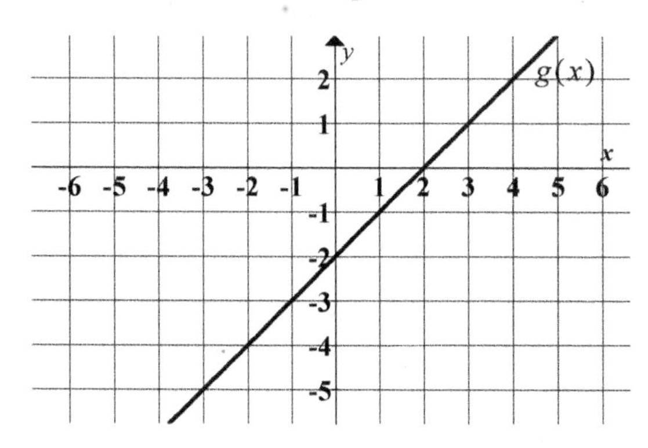

a. Use the graphs to find…

i. $f(-2) =$ undefined

ii. $\lim_{x \to -2} f(x) =$ -4

iii. $g(-2) =$ -4

iv. $\lim_{x \to -2} g(x) =$ -4

b. **True** or **False**: $g(x) = f(x)$. If false, cite a number $x = a$ where $g(a) \neq f(a)$.

x = -2

c. **True** or **False**: $\lim_{x \to a} g(x) = \lim_{x \to a} f(x)$ for all a. If false, cite a number $x = a$ on the graph where $\lim_{x \to a} g(x) \neq \lim_{x \to a} f(x)$.

True

15. There is an important relationship between the functions $f(x) = \frac{x^2 - 4}{x + 2}$ and $g(x) = x - 2$.

a. Factor the numerator and explain how this reveals the relationship between $f(x)$ and the function $g(x) = x - 2$.

b. **True** or **False**: $f(x)$ and $g(x)$ are the same function. Explain your reasoning.

false, f(x) is not defined at x = -2

c. (Check your work) Is your answer to part b consistent with the fact that $f(x) = g(x)$ except at $x = -2$? Explain your reasoning.

yes

d. What is $\lim_{x \to -2} g(x) =$ -4 ?

e. Explain why, without a graph of $f(x) = \frac{x^2 - 4}{x + 2}$ or $g(x) = x - 2$, it is easier to find $\lim_{x \to -2} g(x)$ than to find $\lim_{x \to -2} f(x)$.

no factoring

$\lim_{x \to -2} x - \lim 2$

$-2 - 2 = -4$

f. Based on the graph of $f(x) = \frac{x^2 - 4}{x + 2}$ shown in Question 14, what is $\lim_{x \to -2} f(x) =$ -4 ?

g. **True** or **False**: $\lim_{x \to a} f(x) = \lim_{x \to a} g(x)$ for all a. Explain your reasoning.

true because its the same line

h. (Check your work) The function g in this question is the same g in Question 14. Use the graph of g in in Question 14 to check your answers to parts b, d, and g of this question.

16. For the function $f(x) = \frac{x^2 + x - 6}{x + 3}$, guess the following limit: $\lim_{x \to -3} f(x) =$ -5

Show your work and explain your reasoning.
Hint: Start by factoring the numerator of $f(x)$.

$\frac{(x+3)(x-2)}{x+3}$ $\quad$ $g = x - 2$

$\lim_{x \to -3} \frac{x^2+x-6}{x+3}$ $\quad$ $\frac{\lim x^2+x-6}{\lim x+3}$ $\quad$ $\frac{(\lim x)^2 + \lim x - \lim 6}{\lim x + \lim 3}$

$\frac{9-3-6}{-3+3} = \frac{0}{0}$

17. Summary Box L2.2 should justify your solution to the previous question. If it does not, go back and rework your answer to the previous question using the information in Summary Box L2.2.

Summary Box L2.2: Solving a Limit by Using a Similar Function

Let $f(x)$ be a function that is not defined at $x = a$, and $g(x)$ be a function such that $f(x) = g(x)$ near $x = a$ except at $x = a$. Then $\lim_{x \to a} f(x) = \lim_{x \to a} g(x)$.

Extend Your Understanding Questions (to do in or out of class)

18. Do the following to complete the table below. The first row is done for you.

    a. In Column 2, list all values of x where $f(x)$ does not exist.

    b. In Column 4, find $g(x)$ for each x listed in Column 2 (provided it exists).

f	$f(x)$ undefined at $x = ?$	g	$g(x)$ for values of x listed in Column 2 (if $g(x)$ exists)
$f(x) = \dfrac{x-6}{x^2-36}$	*(list two)* $x = 6$ $x = -6$	$g(x) = \dfrac{1}{x+6}$	$g(6) = \dfrac{1}{12}$ Note: $g(-6)$ does not exist
$f(x) = \dfrac{\left[\dfrac{1}{(4+x)} - \dfrac{1}{4}\right]}{x}$	*(list two)* 0, −4	$g(x) = \dfrac{-1}{16+4x}$	$\frac{-1}{16}$, g(−4) DNE
$f(x) = \dfrac{\sqrt{x+5} - \sqrt{5}}{x}$	0, −5, −∞	$g(x) = \dfrac{1}{\sqrt{x+5} + \sqrt{5}}$	$\frac{1}{2\sqrt{5}}$ −5, DNE

19. For each row in the table above, $f(x) = g(x)$ except at the values of x that listed in Column 2. Use this and the table to evaluate each limit.

a. $\lim_{x \to 6} \left(\dfrac{x-6}{x^2-36} \right) =$ 1/12

b. $\lim_{x \to 0} \left(\dfrac{\sqrt{x+5} - \sqrt{5}}{x} \right) =$ $\frac{1}{2\sqrt{5}}$

c. $\lim_{x \to 0} \dfrac{\left[\dfrac{1}{(4+x)} - \dfrac{1}{4}\right]}{x} =$ $-\frac{1}{16}$

d. $\lim_{x \to -4} \dfrac{\left[\dfrac{1}{(4+x)} - \dfrac{1}{4}\right]}{x} =$ DNE

20. In Questions 16 you likely discovered that, for the function $f(x)=\dfrac{x^2+x-6}{x+3}$, the function $g(x)=x-2$ was useful for finding that the value of $\lim_{x\to -3} f(x)=-5$ (because $f(x)=g(x)$, except at $x=-3$). Use algebra to show the same relationship holds between each pair of functions listed below (reprinted from the table in Question 18).

   a. $f(x)=\dfrac{x-6}{x^2-36}$ and $g(x)=\dfrac{1}{x+6}$

   b. $f(x)=\dfrac{\left[\dfrac{1}{(4+x)}-\dfrac{1}{4}\right]}{x}$ and $g(x)=\dfrac{-1}{16+4x}$

21. In order to calculate limits involving radicals it is sometimes necessary to use a technique called **rationalizing the numerator.** For example, to calculate $\lim_{x\to 3}\dfrac{\sqrt{x+1}-2}{x-3}$, we must calculate the following product. Find this product.

$$\left(\frac{\sqrt{x+1}-2}{x-3}\right)\cdot\left(\frac{\sqrt{x+1}+2}{\sqrt{x+1}+2}\right)=$$

Show your work.

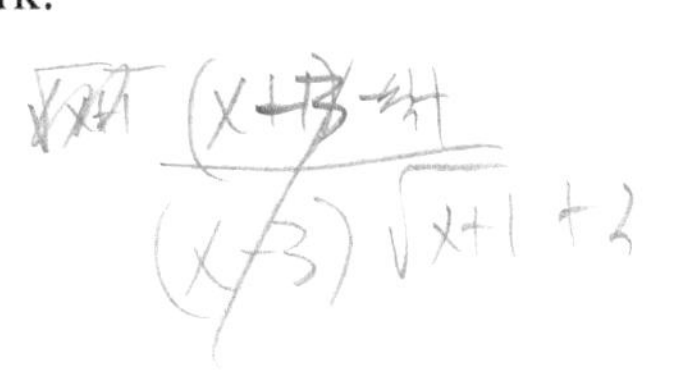

22. (Check your work) Did you find that the function $f(x)=\dfrac{\sqrt{x+1}-2}{x-3}$ is related to the function $g(x)=\dfrac{1}{\sqrt{x+1}+2}$? If not, go back and check your work.

   a. Describe the relationship between $f(x)$ and $g(x)$.

   b. Use the relationship between $f(x)$ and $g(x)$ to calculate $\lim_{x\to 3}\dfrac{\sqrt{x+1}-2}{x-3}=$

23. The solution to which limit in Question 19 can be justified using the technique of rationalizing the numerator? Rationalize the numerator for this function. Show your work.

24. Use the technique in Summary Box L2.2 to evaluate each of the following limits. Show all work.

a. $\lim_{x \to 3} \frac{x^2 - 4x + 3}{x^2 - 9} =$

b. $\lim_{h \to 0} \frac{(3+h)^3 - 27}{h} =$

c. $\lim_{t \to 25} \frac{\sqrt{t} - 5}{t - 25} =$

d. $\lim_{x \to -2} \frac{2x + 4}{16 - 4x^2} =$

e. $\lim_{x \to 2} \frac{\frac{1}{x} - \frac{1}{2}}{x - 2} =$

f. $\lim_{t \to 0} \frac{\sqrt{9 + 2t} - 3}{t} =$

g. $\lim_{x \to 2} \frac{(x+2)^2 - 4x^2}{x - 2} =$

h. $\lim_{\theta \to \frac{\pi}{2}} \frac{\tan \theta}{\sec \theta} =$

i. $\lim_{x \to \frac{\pi}{4}} \frac{\sin x - \cos x}{\tan x - 1} =$

25. (Check your work) The two functions graphed below have the characteristics that…

i. $f(x)$ is un̲defined at $x=a$.

ii. $f(x)=g(x)$ whenever $x \neq a$.

iii. $\lim_{x \to c} g(x) = g(c)$ for all c.

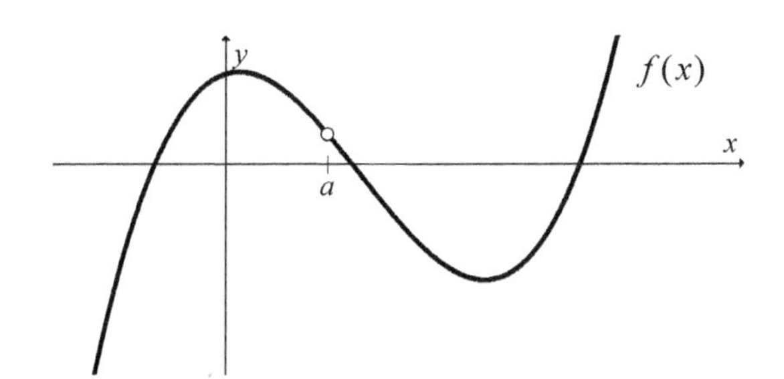

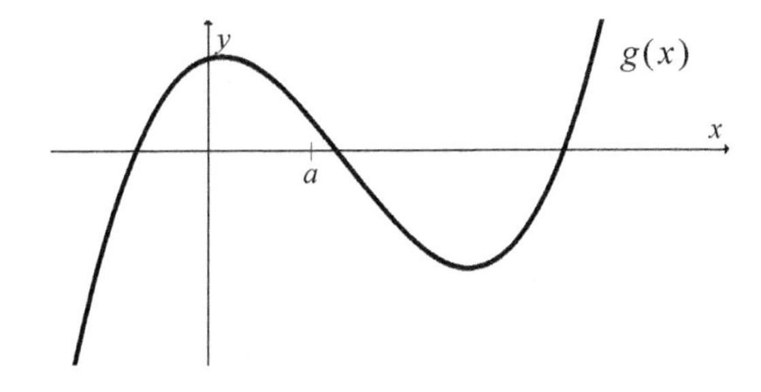

Mark each True or False and briefly explain your reasoning.

a. $\lim_{x \to c} f(x) = f(c)$ for all c. False

b. $\lim_{x \to c} f(x) = g(c)$ for all c. true

c. $\lim_{x \to a} f(x)$ does not exist at $x = a$

false

Notes

L3: Precise Definition of a Limit

Model 1: Error Tolerance

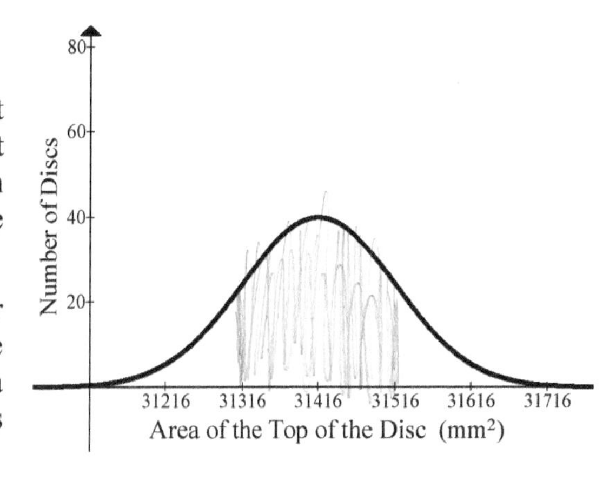

One key reason a car lasts a long time without needing major repairs is that its moving parts fit together well. To ensure this, parts must be made with a **low tolerance for error**. That is, each part has to be the right size, or very close to the right size.

To test the error in a part manufacturing process, a car company uses electrical conductivity to measure the area of the top of a disc-shaped part. The results for a shipment of 10,000 discs are approximated by this model at right.

Construct Your Understanding Questions (to do in class)

1. According to the data in Model 1…
   a. What was the most common disc area measured in the study plotted in Model 1?
   b. (Check your work) The car company specified that each disc should have an area as close as possible to $31{,}416\ \text{mm}^2$. Is this consistent with your answer to part a?

2. The car company will only use parts with a disc area within $100\ \text{mm}^2$ of the target value, or $31{,}416 \pm 100\ \text{mm}^2$. We therefore say that ε, the **upper bound of tolerable error**, is equal to 100. (ε, *epsilon*, is the *Greek* letter **e**, which stands for **e**rror.)
   a. Shade an area on the graph in Model 1 representing the discs that fall within the range represented by $\varepsilon = 100$.
   b. Comment on whether this shipment does a good job of fulfilling the car company's needs. Explain your reasoning.

3. The same conductivity study is done on a different shipment of the same part. The graph at the right approximates the results.
   a. Shade an area on the graph at right representing the discs that fall within the range represented by $\varepsilon = 100$.
   b. Does this shipment represent better or worse manufacturing than the shipment in Model 1? Explain your reasoning.

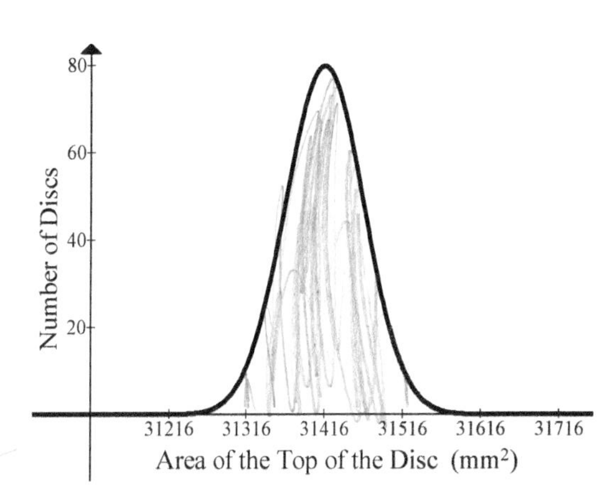

4. It turns out that the part maker manufactures the discs that are the subject of this activity according to *radius*, not area. Given the area, find the radius of each disc.

   a. If $A = 31{,}416\ \text{mm}^2$, then $r =$

   Report answers to three decimal places (i.e., 100.**XXX**)

   b. If $A = 31{,}\mathbf{426}\ \text{mm}^2$, then $r =$

   c. If $A = 31{,}\mathbf{516}\ \text{mm}^2$, then $r =$

5. The part maker must convert $\varepsilon = 100$ to a corresponding **d**ifference in radius, which we will call δ. (*delta* is the *Greek* letter **d**, which stands for **d**ifference).

   a. Demonstrate that two of the calculations in Question 4 (which ones?) can be used to show that $\delta = 0.16$ mm corresponds to $\varepsilon = 100\ \text{mm}^2$.

   b. Now assume that the car company needs to tighten the standards for this part to $\varepsilon = 10\ \text{mm}^2$. Demonstrate that two of the calculations in Questions 4 (which ones?) can be used to show that $\delta = 0.016$ mm corresponds to the car company's <u>**new**</u>, tighter error boundary of $\varepsilon = 10\ \text{mm}^2$.

   c. Based on the pattern above, if the car company tightens its standards to $\varepsilon = 1\ \text{mm}^2$, what would be the corresponding value of δ?

6. The function $A(r) = \pi r^2$, graphed at right, relates area to radius for a circular disc.

   a. Estimate the y (vertical axis) coordinate of the point marked on the graph.

   b. Estimate the r (horizontal axis) coordinate of the point marked on the graph.

   c. (Check your work) Are your answers to parts a and b consistent with the fact that the point marked on the graph represents the ideal area and radius of the car part we have been discussing in this activity?

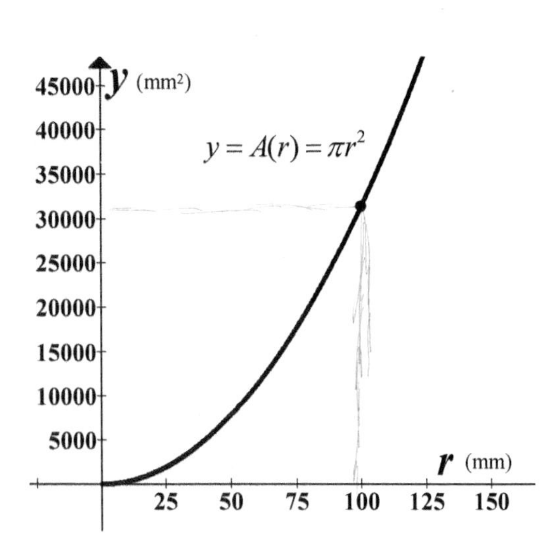

7. The graph at right shows a zoom view of the point marked on the graph in the previous question.

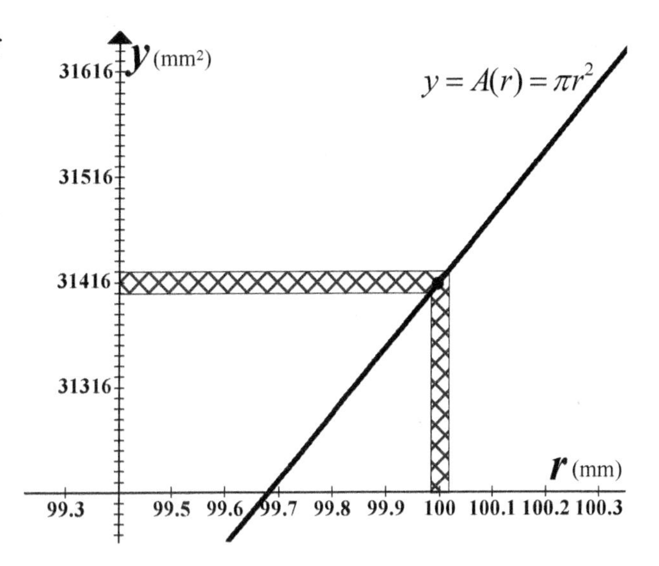

   a. The portion of $y = A(r)$ that passes through the shaded region on the graph represents the acceptable values of r and $A(r)$ for an error tolerance boundary of $\varepsilon =$ ________

   b. Use the graph to make a rough estimate of $\delta =$ ________

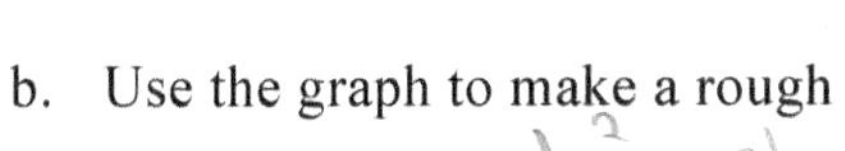

8. (Check your work) Your estimates for ε and δ in the previous question should be close to one set of values given in Question 5. Which set in Question 5 do they match most closely:

   5a, **5b**, or **5c** ? [circle one] If none match, check your answers to Question 7.

9. Let the target area and radius of the disc be L and c, respectively, as marked.

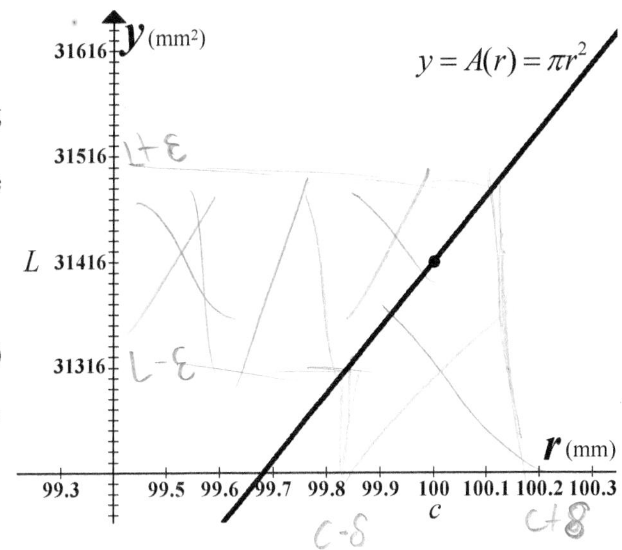

   a. For $\varepsilon = 100$, label the value along the vertical (y) axis that is equal to "$L + \varepsilon$." Then label the value that is equal to "$L - \varepsilon$."

   b. Shade an area like the one on the graph above, but for $\varepsilon = 100$.

   c. Make a mark on the horizontal (r) axis and label it "$c + \delta$." Make another mark and label it "$c - \delta$."

   d. Use the graph to estimate

      $\delta =$ ________

   e. (Check your work) This estimate is closest to δ given in Question **5a**, **5b**, or **5c**. If it is not close to any of these, go back and check your work. [circle one]

10. Describe in words what happens to δ as ε gets smaller. For example, as ε goes from 100 to 10 to 1, as we imagined in Questions 5a, 5b, and 5c, respectively.

11. In a manufacturing process there is a bound to how precisely a part can be made. If the car company asks for smaller and smaller values of ε, the part maker will eventually have to reply that their factory cannot deliver the part to such tight specifications because it is smaller than their minimum achievable δ.

    Is this the case for the function $A(r) = \pi r^2$, shown on the graphs on the previous page?

    That is, which one of the following statements is **TRUE**?

    a. If we choose a value of ε small enough then it will be impossible to find a corresponding δ that is sufficiently small.

    b. No matter how small we choose ε there exists a corresponding value of δ.

    Circle a or b and explain your reasoning.

12. For $A(r)$, r, L, c, ε and δ as previously described in this activity.

    a. Which limit is true?

    $\lim_{r \to 0} A(r) = L$ $\qquad$ $\lim_{r \to 0} A(r) = c$ $\qquad$ $\lim_{r \to c} A(r) = L$

    Circle one and explain your reasoning.

    limit as radius approaches C of the area IS L

    b. Which best describes the condition set by the car company for the disc?

    $L < \varepsilon$ $\qquad$ $A(r) < \varepsilon$ $\qquad$ $|A(r) - L| < \varepsilon$ $\qquad$ $|A(r) - c| < \varepsilon$ $\qquad$ $A(r) + c > \varepsilon$ $\qquad$ $|A(r) - L| > \varepsilon$

    Circle one and explain your reasoning.

    c. Which best describes the condition used by the part maker to manufacture the disc?

    $r < \delta$ $\qquad$ $c > \delta$ $\qquad$ $|r - c| < \delta$ $\qquad$ $r + c > \delta$ $\qquad$ $|r - L| < \delta$

    Circle one and explain your reasoning.

13. The successively tighter error tolerances in Question 5 are reminiscent of the informal definition of a limit at the end of CalcActivity L1. In Summary Box L3.1 we replace that definition with one that is more precise.

    a. Describe in words what ε represents in this definition.

    b. Describe in words what δ represents in this definition.

Summary Box L3.1 Precise Definition of a Limit

Let c be a number in an open interval, and f be a function defined at every number in that interval, except perhaps at c. The $\lim_{x \to c} f(x) = L$ if...

for any $\varepsilon > 0$, there exists a $\delta > 0$ such that $|f(x) - L| < \varepsilon$ whenever $0 < |x - c| < \delta$.

14. Which endings represent a valid restatement of the Precise Definition of a Limit in Summary Box L3.1? **Circle all that apply**, and write down any questions that you have for each.

    The limit of $f(x)$ as x approaches c is equal to L if...

    a. for any positive value of ε, no matter how small, there exists a sufficiently small, positive distance δ such that $f(x)$ is within ε of L whenever x is within δ of c (but not equal to c).

    b. ε, the distance from $f(x)$ to L, can be made as small as we like by choosing δ, the distance from x to c ($x \neq c$), sufficiently small.

    c. for every $\varepsilon > 0$ we can find $\delta > 0$ such that $f(x)$ is found in the interval $(L - \varepsilon, L + \varepsilon)$ whenever x lies in the interval $(c - \delta, c + \delta)$, but is not equal to c.

Extend Your Understanding Questions (to do in or out of class)

15. (Check your work) All three endings in the previous question are valid. Which one, if any, made the most sense to you? Explain why.

16. Consider $f(x) = \dfrac{2x^2 - 6x + 4}{x - 2}$, shown on the graph.

    Does $\lim_{x \to 2} f(x)$ exist? Explain your reasoning.

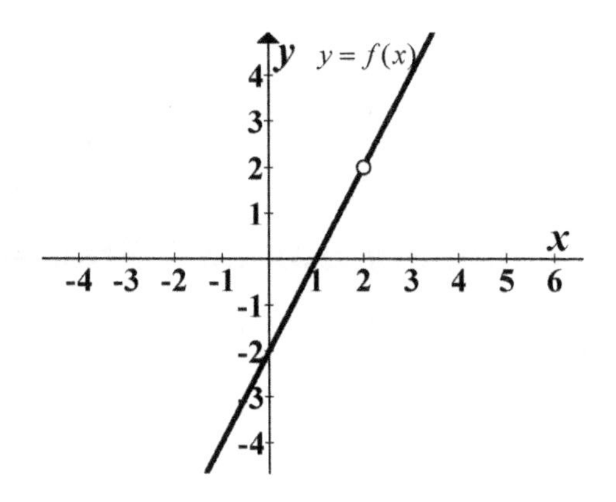

17. Levi uses the following reasoning to show that, for the function in the previous question, $\lim_{x \to c} f(x) = L$ when $L = 2$ and $c = 2$. That is, $\lim_{x \to 2} f(x) = 2$.

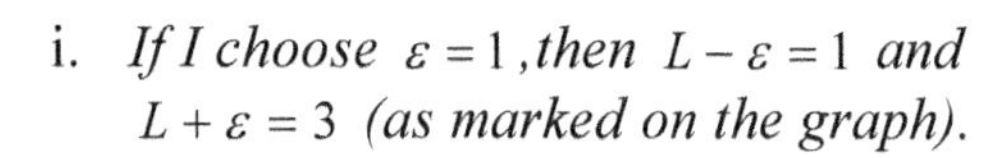

    i. *If I choose* $\varepsilon = 1$*, then* $L - \varepsilon = 1$ *and* $L + \varepsilon = 3$ *(as marked on the graph).*

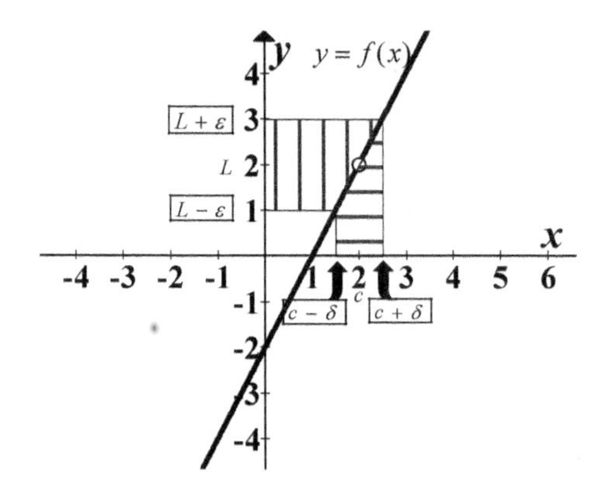

    ii. *Now I find the corresponding positive value of* δ *such that whenever* x *is in the interval* $(c - \delta, c + \delta)$*, but not equal to 2,* $f(x)$ *is in* $(L - \varepsilon, L + \varepsilon)$*. The answer is* $\delta = 0.5$*, as shown on the graph.*

    iii. *Does this work if I choose* $\varepsilon = 0.1$*? The answer is yes, with* $\delta = 0.05$*.*

    Does this work if I choose $\varepsilon = 0.01$*? The answer is yes, with* $\delta = 0.005$*.*

    In fact, for this function it seems that $\delta = 0.5\varepsilon$ *for any positive* δ*.*

    iv. *Since I can find a satisfactory* δ *for any positive value of* ε*, I conclude that* $\lim_{x \to 2} f(x) = 2$*.*

    Is Levi's proof valid? Explain why or why not.

18. (Check your work) Rosa says that while Levi has made some very nice observations, using the words "it seems that" in a proof has made her conclude that more work needs to be done. She turns his ideas into a more formal argument. Fill in the blanks of her proof.

Let $\varepsilon > 0$ be given. We need to find δ such that if $0 < |x-2| < \delta$, then $|f(x)-2| < \varepsilon$. The function $f(x) = \dfrac{2x^2-6x+4}{x-2}$ can be written as $\dfrac{2x^2-6x+4}{x-2} = \dfrac{2(\_\_\_\_\_\_\_\_)}{x-2} = \dfrac{2(x-2)(x-1)}{x-2}$

and since we are only considering values of x that are **not equal** to ____, this is equivalent to $2(x-1)$. Suppose that $\frac{\varepsilon}{2} < 1$, since $0 < |x-2| < \frac{\varepsilon}{2}$, we have $|x-2| < \frac{\varepsilon}{2}$. In other words,

$-\frac{\varepsilon}{2} < x-2 <$ ________ . Multiplying this inequality by ____________ will give

$-\varepsilon < 2(x-2) < \varepsilon$, or $-\varepsilon < 2(x-1)-2 < \varepsilon$. This will give us $|2(x-1)-2| <$ ______, our desired

conclusion. Now if $\frac{\varepsilon}{2} > 1$, then $\varepsilon > 2$. If $0 < |x-2| < 1$ then multiplying again by _____ will give us $|2(x-1)-2| < 2 < \varepsilon$.

19. Consider the function f shown on the graph at right.

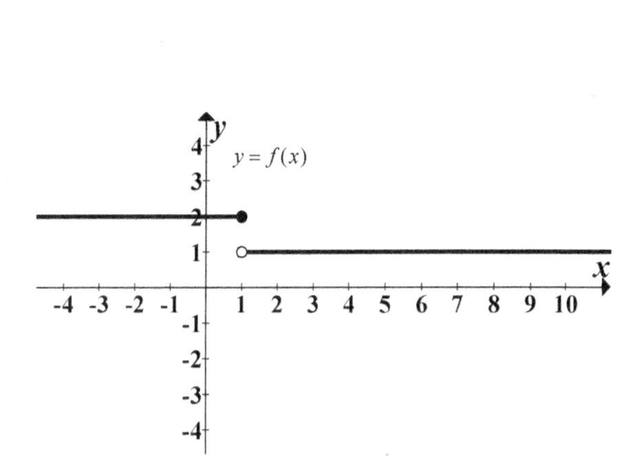

a. Does $\lim_{x\to 1} f(x)$ exist? Explain your reasoning.

b. A student says $\lim_{x\to 1} f(x) = 2$. Prove that this is incorrect by finding a value of ε that is ***so small*** that we cannot find a corresponding number δ that is ***small enough*** to satisfy the definition in Summary Box L3.1. In other words, find a value of ε for which we cannot find δ small enough so that for **every** x in the interval $(1-\delta, 1+\delta)$ (except $x=1$), we make $f(x)$ lie in the interval $(2-\varepsilon, 2+\varepsilon)$.

20. (Check your work) Is your answer to the previous question consistent with the fact that any value of ε between zero and 1 can be used to show that $\lim_{x\to 1} f(x) \neq 2$?

Notes

Limits 4: Continuity

Model 1: Continuity

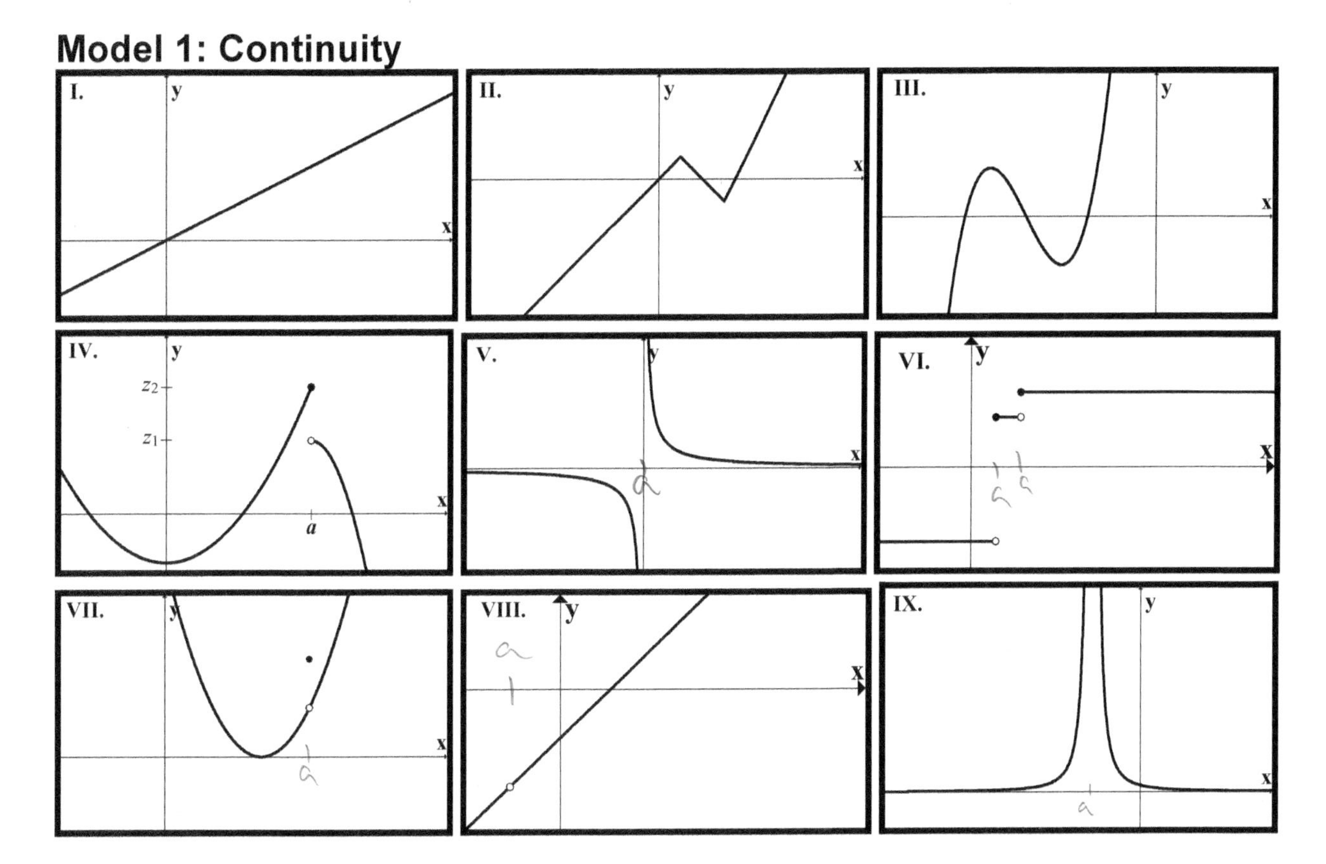

Construct Your Understanding Questions (to do in class)

1. Which is the correct value of $f(a)$ in **Graph IV** in Model 1: z_1 or z_2? Circle one and explain your reasoning. (Hint: Recall the meaning of an open versus a filled circle.)

2. In Model 1, only the functions in **Graphs I-III** are **continuous** over the interval shown.

   a. Each of the other functions (shown on **Graphs IV-IX**) is **discontinuous** at one or more points. Mark each point of discontinuity by adding a tick mark and a letter "a" along the x axis as shown on **Graph IV**. If there are multiple points of discontinuity, mark them a_1, a_2, etc.

   b. Based on the information in Model 1, make up a definition for the term **continuous function** (on an interval).

3. (Check your work) A crude test of continuity is to imagine an infinitely small ant walking along the graph of a function. If the ant can travel along the curve without interruption (e.g. without falling in a hole such as the one in Graph VII in Model 1) the function is continuous over that interval. Is your definition consistent with this crude test? Explain.

4. List each graph in Model 1 where $\lim_{x\to a^-} f(x) \neq \lim_{x\to a^+} f(x)$ for some point a on the interval shown.

   a. **True** or **False**: If $\lim_{x\to a^-} f(x) \neq \lim_{x\to a^+} f(x)$ then f is discontinuous at a.

      If false, draw a function that is <u>continuous</u> at a but $\lim_{x\to a^-} f(x) \neq \lim_{x\to a^+} f(x)$.

   b. (Check your work) It is a common error to assume that corners, such as those in **Graph II** in Model 1, are points of discontinuity. This leads some students to erroneously conclude that part a of this question is false. A graph with corners (e.g. **Graph II** in Model 1) is <u>continuous</u>, and part a of this question is true.

   c. Recall that ∞ is not a real number, so if $\lim_{x\to a} f(x) = \infty$ then this limit does not exist.

      i. Which graph in Model 1 has a point a where $\lim_{x\to a^-} f(x) = \lim_{x\to a^+} f(x) = \infty$?

      G

      ii. Is this graph continuous at a? Explain your reasoning.

      no

      lim f(x) DNE

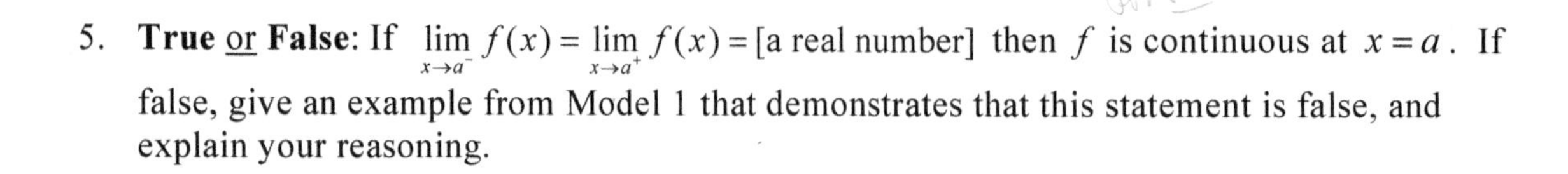

5. **True** <u>or</u> **False**: If $\lim_{x\to a^-} f(x) = \lim_{x\to a^+} f(x) =$ [a real number] then f is continuous at $x = a$. If false, give an example from Model 1 that demonstrates that this statement is false, and explain your reasoning.

6. (Check your work) There are two graphs in Model 1 that demonstrate the statement in the previous question is false.

   a. Identify the Roman numerals of these two graphs.

   b. For each of these two graphs, write down what you know about the value of $f(a)$.

   c. Each of these two graphs has a "problem" at $x = a$ that makes the function discontinuous at a. Describe how these two problems are different from one another.

   d. **Graphs VII and VIII** each have a **point** (or **removable**) **discontinuity**. In a way, each of these functions is discontinuous at a because $f(a)$ is <u>not</u> the value we *expect* based on the function near a. What would *fix* this? That is, the graph would be continuous at a if $f(a) =$

   Choose from: a $f(a)$ $\lim_{x \to a} f(x)$ $f(x)$ x **and explain your reasoning.**

7. (Check your work) What part of Summary Box L4.1 confirms your answer to part d of the previous question? If it does not confirm your answer, reconsider your choice, above.

Summary Box L4.1: Definitions of Continuity At a Point and On an Interval

A function f is continuous at $x = a$ if $\lim_{x \to a} f(x) = f(a)$.

A function f is continuous on an **interval** (c,d), if it is continuous at every point in the interval.

If this interval is **closed** $[c,d]$, then it must be also be true that...

- at c, the left endpoint, $\lim_{x \to c^+} f(x) = f(c)$, i.e. $f(x)$ is "continuous from the right" at $x = c$;
- at d, the right endpoint, $\lim_{x \to d^-} f(x) = f(d)$, i.e., $f(x)$ is "continuous from the left" at $x = d$.

8. With your group, come up with a real world example of a process described by a continuous function, and one described by a function that contains one or more discontinuities. Sketch a graph of each of your functions, and *be prepared to report your answer as part of a whole-class discussion.*

Model 2: Identifying a Discontinuity from an Equation

I. $f(x)=\frac{1}{2}x$	**II.** $f(x)=\begin{cases} x & \text{if } x\le 1 \\ -x+2 & \text{if } 1<x\le 3 \\ 2x-7 & \text{if } x>3 \end{cases}$	**III.** $f(x)=x^3+9x^2+25x+21$
IV. $f(x)=\begin{cases} \frac{1}{2}x^2-2 & \text{if } x\le 6 \\ -x^2+12x-33 & \text{if } x>6 \end{cases}$	**V.** $f(x)=\frac{1}{x}$	**VI.** $f(x)=\begin{cases} -3 & \text{if } x<1 \\ 2 & \text{if } 1\le x<2 \\ 3 & \text{if } x\ge 2 \end{cases}$
VII. $f(x)=\begin{cases} \frac{1}{2}x^2-4x+8 & \text{if } x\ne 6 \\ 4 & \text{if } x=6 \end{cases}$	**VIII.** $f(x)=\frac{x^2-4}{x+2}$	**IX.** $f(x)=\frac{1}{(x+2)^2}$

Extend Your Understanding Questions (to do in or out of class)

9. At what value of x does $f(x)$ have a discontinuity for…

   a. **Function V** in Model 2?

   b. **Function VIII** in Model 2?

   Explain your reasoning.

10. For **Function VI** in Model 2 (a piece-wise defined function), what is…

    a. $f(1)=$

    b. $\lim_{x\to 1^+} f(x)=$

    c. $\lim_{x\to 1^-} f(x)=$

    d. $f(2)=$

    e. $\lim_{x\to 2^+} f(x)=$

    f. $\lim_{x\to 2^-} f(x)=$

    Is **Function VI** continuous? Use your answers to parts a-f to justify your answer.

11. In the previous question, we checked **Function VI** for continuity at $x=1$ and $x=2$.

    a. What two values of x should be checked for continuity for **Function II** in Model 2?

    b. Is **Function II** in Model 2 continuous? Show your work.

    c. What value(s) of x should be checked for continuity for **Function IV** in Model 2?

    d. Is **Function IV** in Model 2 continuous? Show your work.

12. Use what you discovered above to mark **Functions I, III, VII,** and **IX** as continuous or not continuous (and note any values of x where a function has a discontinuity).

13. (Check your work) The functions in Model 2 are the same functions that appear as graphs in Model 1. Use this to check your answers to the previous question, and correct any answers.

14. Based on what you know so far, sort the six discontinuous functions (IV-IX) in Models 1 and 2 into the three categories listed on the table below. (Put each Roman numeral in exactly one box in the column labeled **Roman numerals**..., below.)

Type of Discontinuity	**Roman numerals** of functions in Models 1 and 2 that have a..	**Description of a graph** of a function that has a...	**Characteristics of an equation** that should be checked for a...
Jump			
Infinite			
Point			

15. In the column labeled **Description of a graph** (above), describe the key defining characteristics of each type of discontinuity listed.

16. In the column labeled **Characteristics of an equation** (above), describe what you look for in an equation that may indicate the presence of each type of discontinuity listed.

17. (Check your work) Is your answer to the previous question consistent with Summary Box L4.2? Note any differences and discuss them with your group.

Summary Box L4.2: Some Properties of Continuous Functions

Piecewise defined functions must be checked for continuity, and rational functions are discontinuous when the denominator is equal to zero; however, there are other functions that are discontinuous, for example, $f(x) = \tan x$.

If two functions f and g are continuous at $x = a$, then all of the following are also continuous at a:

$f+g$, $f-g$, cf ($c =$ constant), fg, $\frac{f}{g}$ $(g \neq 0)$, and $f \circ g$ (f continuous at $g(a)$)

18. Circle the two statements below that are true. For the remaining three false statements, sketch a graph of a function demonstrating that it is false.

    a. If $\lim_{x\to a} f(x) = f(a)$ then f is continuous at $x = a$

    b. If f is not continuous at $x = a$ then $f(a) \neq \lim_{x\to a} f(x)$.

    c. If f is not continuous at $x = a$ then $f(a)$ is undefined.

    d. If f is not continuous at $x = a$ then $\lim_{x\to a} f(x)$ does not exist.

    e. If f is not continuous at $x = a$ then either $f(a)$ is undefined or $\lim_{x\to a} f(x)$ does not exist.

Confirm Your Understanding Questions (to do at home)

19. Consider the function $f(x)=\dfrac{x^2+6x+9}{x+3}$

   a. Is this function continuous? If not, at what value(s) of x is it discontinuous? Explain your reasoning.

   b. The numerator of this function can be factored and the function can be expressed as $f(x)=\dfrac{(x+3)^2}{x+3}$. Note that this is not the same as the function $f(x)=x+3$. Construct an explanation for why and how these two very similar functions differ from one another.

   c. Sketch a graph of these two functions.

20. On what interval is the function $f(x)=\sqrt{2-x^2}$ continuous? Explain your reasoning.

21. For **Graph IV** in Model 1, the $\lim_{x\to a^-} f(x)=f(a)$.

   a. (Review) Explain the significance of the superscript "–" in the equation above.

   b. Use the term $\lim_{x\to a^+} f(x)$ in an explanation for why the function shown on **Graph IV** is not continuous at a.

Notes

Derivatives 1: Velocity, Introduction to Derivatives

Model 1: Driving to a Cabin in the Woods

A family drives from their home in the city to a cabin in the woods. Before they begin, they reset the trip odometer in their car. Then, every hour they record the trip odometer reading (the total distance they have traveled). The following graph shows the results.

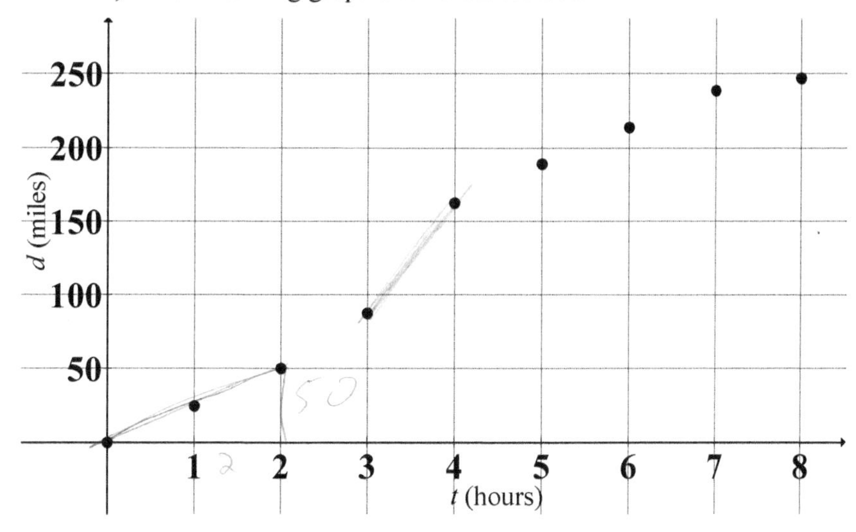

Construct Your Understanding Questions (to do in class)

1. Draw a line on the graph in Model 1 from the point at $t_0 = 0$ to the point at $t_1 = 2$.

   a. What is the slope (m) of this line? **Include units in your answer.**

   b. Describe in words what this slope (m) tells you about the movement of the car during this time period.

2. On the graph in Model 1, draw a line segment between the points that mark the one hour during which the car was likely traveling on an interstate highway, and explain your reasoning.

3. In general, what information is conveyed by the slope (m) of a line between any two points on a distance versus time function?

Model 2: Car Rolling Down a Ramp

A car starts at rest at $t = 0$. Assume the distance (d) the car travels is always a function of time (t) according to the equation: $d(t) = 3t^2$ with units: d (in meters); t (in seconds). The solid curve below shows $d(t) = 3t^2$.

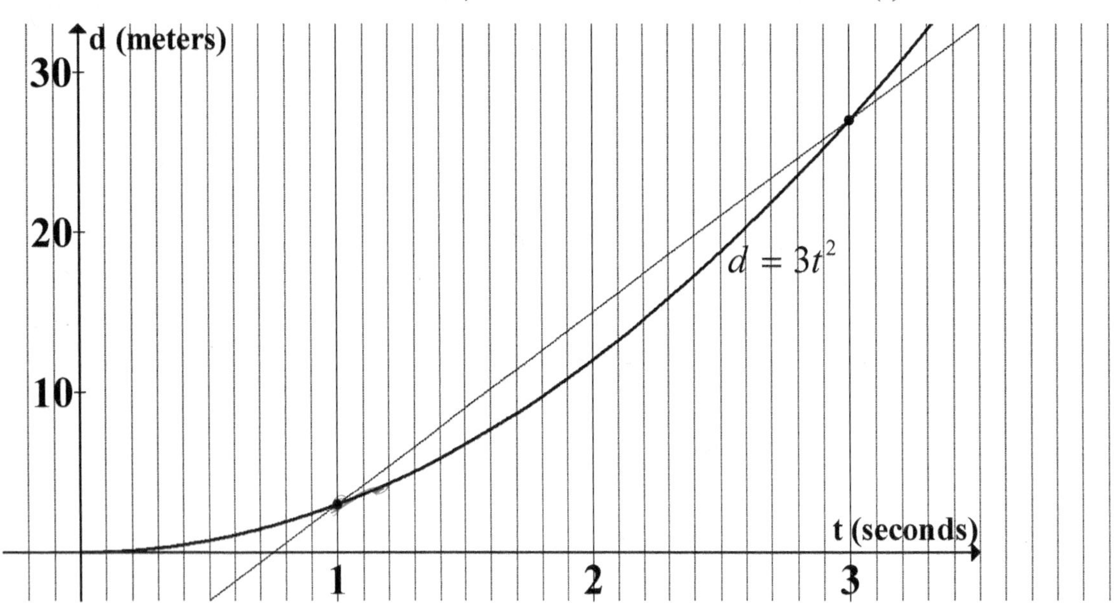

Construct Your Understanding Questions (to do in class)

4. (Check your work) The slope (m) of the straight line in Model 2 tells you the **average** velocity of the car between $t = 1$ and $t = 3$ seconds. Is this consistent with your answer to the previous question? If not, discuss with your group and resolve any differences. yes

5. Circle the formula that gives the average velocity (v_{ave}) of the car in Model 2 between a time of interest, which we will call a, and a later time, t.

$$v_{ave} = 3t^2 - 3a^2 \qquad v_{ave} = \frac{3t^2 - 3a^2}{t - a} \qquad v_{ave} = \frac{3t^2}{t}$$

6. (Check your work) Use the formula you chose to confirm that the average velocity of the car in Model 2 is 12 m/s if you set $a = 1$ and $t = 3$. Show your work.

7. Consider the points on the curve in Model 2 at $(1,3)$ and $(1.2, 4.32)$.

   a. Add a straight line to the graph in Model 2 that passes through these two points. (Note: The slope of this line is 6.6 m/s.)

   b. Let's think of the instantaneous velocity of the car at time t as being the speedometer reading at time t. Which slope (**12 m/s** or **6.6 m/s**) is a better estimate of instantaneous velocity at our time of interest, $a = 1$ second? Explain your reasoning.

8. A student says that using $a = 1$, and $t = 1.1$ in the formula $v_{ave} = \frac{3t^2 - 3a^2}{t - a}$ generates an estimate of the instantaneous velocity of the car at $a = 1$ second that is even better than 6.6 m/s.

   a. Do you **agree** or **disagree**? Circle one and explain your reasoning.

   yes

   b. Choose a value of t that would give an even better estimate of the instantaneous velocity of the car at $a = 1$ second.

   1.001

   c. Does your answer in part b give the very best estimate? If not, describe in words how you can use the formula at the top of the page to generate increasingly better estimates for the instantaneous velocity of the car at $a = 1$ second.

   bring t and a closer

9. What is the value of the function $v_{ave} = \frac{3t^2 - 3a^2}{t - a}$ when $a = t$ (e.g., if $a = 1$ and $t = 1$)?

   undefined

10. Write down a limit that gives the instantaneous velocity of the car at a seconds.

$v_{inst} = \lim_{t \to a} \frac{3t^2 - 3a^2}{t - a}$

11. Instantaneous velocity at $a = 1$ second can also be represented graphically.

   a. Sketch the line on the graph in Model 2 whose slope (m) is equal to the instantaneous velocity of the car at $a = 1$ second.

   b. Since it is hard to draw this line accurately (and it is very close to the line you drew representing the average velocity between $a = 1$ and $t = 1.2$), briefly describe the characteristics of this line representing the instantaneous velocity at $a = 1$ second.

   tangent

12. (Check your work) Is your answer to the previous question consistent with the information in Summary Boxes D1.1 and D1.2? If not, go back and revise your answer to the previous question.

Summary Box D1.1: Definition of the Tangent Line to a Curve at a Point

On a distance versus time curve, a line **tangent** to the curve at a point P on the curve, is a line that passes through P and has a slope equal to the instantaneous velocity at P.

More generally, a tangent line to a curve $y = f(x)$ at a point P given by the coordinates $(a, f(a))$ is the line through P with a slope equal to…

$$m = \lim_{x \to a}\left[\frac{f(x) - f(a)}{x - a}\right]$$

Summary Box D1.2: Instantaneous Rate of Change (the Derivative)

Velocity is one example of a rate of change. In this activity we saw that the instantaneous rate of change at a point is equal to the slope of the tangent line to the position function at that point.

Such instantaneous rates of change are so important in science and mathematics that they are given a special name, a **derivative**. The derivative of a function f at $x = a$ is usually written as $f'(a)$ and is defined as $f'(a) = \lim_{x \to a}\left[\frac{f(x) - f(a)}{x - a}\right]$

13. A function $g(t)$ gives the position of a moving object as a function of time. Describe what each of the following tell you about the movement of the object at $t = a$ (include possible units):

    a. $g(a)$ Position

    b. $g'(a)$ Speed

14. Let $g(t) = 3t^2$. Use the formula in Summary Box D1.2 to determine the slope of the tangent line to the curve at $a = 2$. Show your work. (Hint: factor the numerator.)

Extend Your Understanding Questions (to do in or out of class)

15. (Check your work) Use the graph of $g(t)=3t^2$ in Model 2 to check if your answer to the previous question is reasonable.

16. Find the slope of the tangent line to $f(x)$ at the value of x indicated. Show your work.

    a. $f(x)=x^2$ at $x=2$

    b. $f(x)=\frac{1}{x}$ at $x=4$

    c. $f(x)=-x^2+9x$ at $x=1$

    d. $f(x)=4x^2-2x+3$ at $x=0$

    e. $f(x)=\sqrt{x}$ at $x=4$

17. For each function, find the equation for $f'(a)$. [Your answer will include a terms.] Show your work.

    a. $f(x)=x^2$

    b. $f(x)=\frac{1}{x}$

18. (Check your work) Use the general expressions you generated in Question 17 to check your answers to parts a and b of Question 16.

Notes

Derivatives 2: The Derivative at a Point

Model 1: Review of Velocity

In the previous activity we explored position functions (distance versus time) and learned how to find **average velocity** between two points and **instantaneous velocity** at a point. Some key conclusions were:

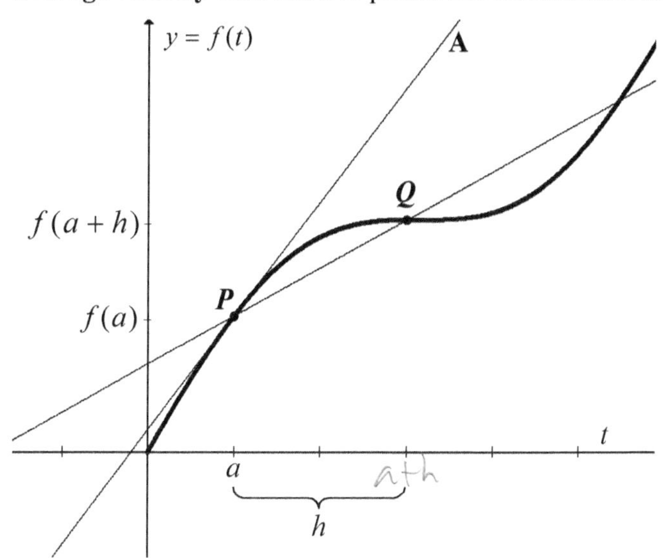

- The slope of a line through a point P and another point Q on the graph of a position function f is equal to the average velocity over this interval. (We call this type of line a **secant line**.)
- The slope of the line tangent to the graph of f at P is equal to the instantaneous velocity at P. The slope of this **tangent line** is given by a special limit called a **derivative**.

Construct Your Understanding Questions (to do in class)

1. If $y = f(t)$, the solid curve in Model 1, shows position versus time for a car, then…
   a. What does the slope of the line marked **A** tell you about the car? Explain.
   b. At which point is the car going faster ***P*** or ***Q***? (circle one)
   c. The quantity h is shown as a distance along the horizontal axis. Write the label $a+h$ below the hash mark that represents the time, $t = a+h$.
   d. Write an expression for the slope of the secant line that goes through P and Q. (Hint: First write down the coordinates of the points P and Q.) According to Model 1, what does this slope tell you about the movement of the car?

$m_{PQ} =$ ____________ m_{PQ} tells you…

2. Look at the graph in Model 1 and imagine h getting smaller and smaller, causing Q to approach P. When h is infinitesimally small what does the quantity m_{PQ} tell you about the movement of the car? Explain your reasoning.

3. Use your equation for m_{PQ} to generate an expression for the quantity in Question 2.

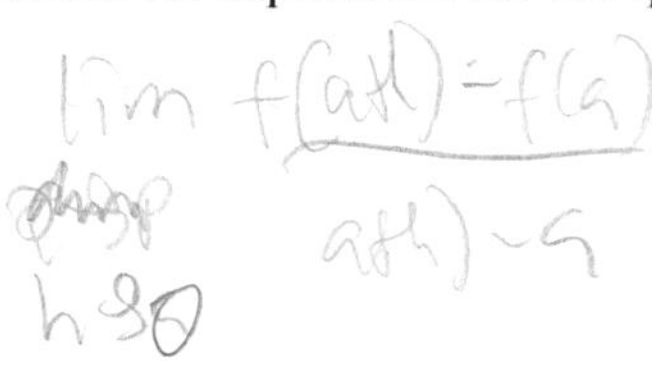

4. (Check your work) Does your answer Question 3 match Summary Box D2.1? If not, revise your answer to Question 3 or show it is equivalent to the equation below.

Summary Box D2.1: Definition of the Derivative at a Point

For a function f, the derivative at a point, $f'(a)$, can be expressed as...

$$f'(a) = \lim_{h \to 0} \frac{f(a+h) - f(a)}{h}$$

5. Use Summary Box D2.1 to find $f'(2)$ for the function, $f(x) = x^2$. Show your work.

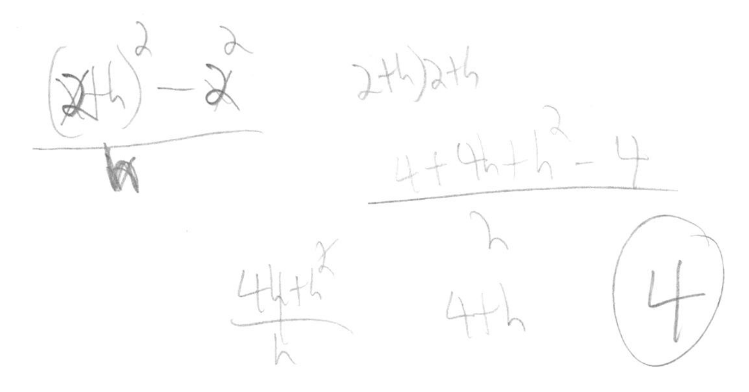

6. Use the formula $f'(a) = \lim_{x \to a} \left[\frac{f(x) - f(a)}{x - a} \right]$ (<u>not the one in Summary Box D2.1</u>) to find $f'(2)$ for $f(x) = x^2$. Show your work.

7. (Check your work) The formula in Question 6 (from Activity D1) is equivalent to the formula in Summary Box D2.1. Are your answers to Questions 5 and 6 the same? If not, go back and check your work.

8. The graph at right shows a heating function, $H(t)$, for a test of a pottery oven.

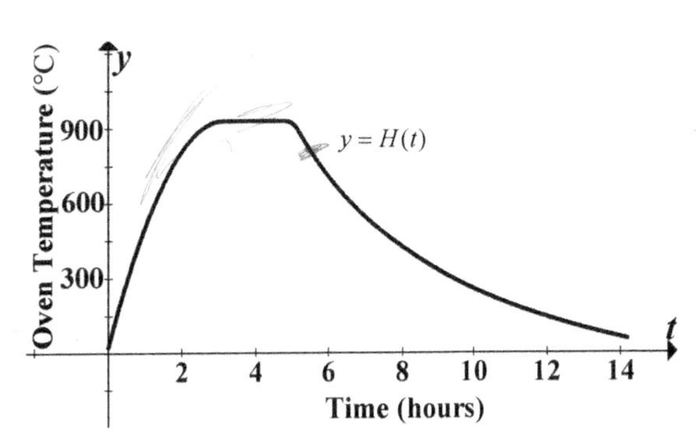

   a. How many hours does it take to heat the oven from room temperature (25 °C) at $t = 0$ to its maximum temperature of 925 °C? (Round your answer to the nearest hour.)

   3

   b. What is the average temperature change per hour during oven heating?

   300

   c. Is $H'(2)$ positive (+), negative (−), or zero (0)? What does this tell you about the oven at $t = 2$?

   pos

   d. Is $H'(4)$ positive (+), negative (−), or zero (0)? What does this tell you about the oven at $t = 4$?

   0

   e. Is $H'(8)$ positive (+), negative (−), or zero (0)? What does this tell you about the oven at $t = 8$?

   neg

   f. Mark the point on the graph where the oven is cooling most rapidly, and explain your reasoning.

   g. In general, at a point $t = a$, what information about the oven is conveyed by the derivative of the heating function at this point, $H'(a)$?

9. Complete the following statements by filling in the blanks. (Check your work) Are these completed statements consistent with your answer to part g of the previous question?

   On a graph of a distance (y) versus time (t) function, the derivative of the function at a point gives the ____Instant____ velocity of the object at that point.

   Given a graph of a temperature (y) versus time (t) on the heating function above, the derivative of the function at a point gives the ____Instant____ rate of temperature change with respect to time of the oven at that point.

10. Give another example of a derivative (an instantaneous rate of change) you might encounter in the real world. Identify the independent variable and the dependent variable, and assign reasonable units to both of these and the derivative. It may help to sketch a graph. Be prepared to report your group's answer as part of a whole-class discussion. **Challenge: Try to find an example that does not use time as one of your variables.**

Model 2: Alternate Notations

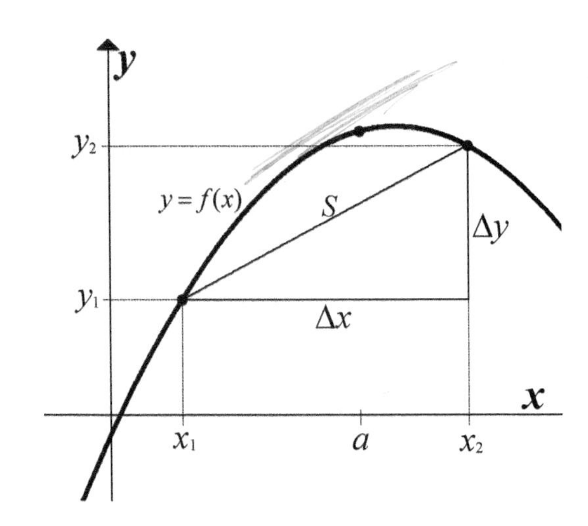

Recall that the <u>average</u> rate of change in y with respect to x between x_1 and x_2 is equal to [the slope of secant line marked with an S] $= \dfrac{y_2 - y_1}{x_2 - x_1} = \dfrac{\Delta y}{\Delta x}$

By analogy to the last of these, we represent the *instantaneous* rate of change in y with respect to x at $x = a$ using the symbols: $\left.\dfrac{dy}{dx}\right|_{x=a}$

This is called Leibniz notation.

The following are equivalent ways of representing $f'(a)$, the derivative of a function $f(x)$ at $x = a$ (assuming the dependent variable is y):

$$\left.\frac{dy}{dx}\right|_{x=a} = y'\big|_{x=a} = \left.\frac{df}{dx}\right|_{x=a}$$

Construct Your Understanding Questions (to do in class)

11. On the graph in Model 2, sketch the tangent line to $y = f(x)$ at $x = a$.

12. Express the following using Leibniz notation:

    a. The slope of the tangent line you drew in the previous question.

    $\left.\dfrac{dy}{dx}\right|_{f(x+h}$

    b. $\lim\limits_{\Delta x \to 0} \dfrac{\Delta y}{\Delta x}$ (at the point where $x = x_1$)

13. For a function $y = f(x)$ that is continuous at *a*, circle the letter of each of the following that is equivalent to $f'(a)$.

a. The slope of the line tangent to the graph of $y = f(x)$ at $x = a$

b. $\lim_{x \to a} f(x)$

c. $\lim_{x \to 0} f(x)$

d. $\lim_{h \to 0} f(x)$

e. The derivative of f at $x = a$

f. $\lim_{x \to a} \left[\frac{f(x) - f(a)}{x - a} \right]$

g. The instantaneous rate of change of *y* with respect to *x* at $x = a$

h. The instantaneous rate of change of *x* with respect to *y* at $x = a$

i. The instantaneous velocity at $x = a$ (where $x =$ time, and $f(x) =$ distance)

j. The instantaneous rate of temperature change at $x = a$ degrees Celsius (where *x* is time in hours, and $f(x)$ is temperature in degrees Celsius)

k. $\lim_{h \to 0} \left[\frac{f(a+h) - f(a)}{h} \right]$

l. $\left. \frac{dy}{dx} \right|_{x=a}$

m. $\left. \frac{df}{dx} \right|_{x=a}$

n. $\left. y' \right|_{x=a}$

o. The slope of a line that passes through the point $(a, f(a))$ but does not intersect the graph of the function at any other point

Extend Your Understanding Questions (to do in or out of class)

14. (Check your work) The last statement in the previous question illustrates two common **misconceptions** about tangent lines. Write "FALSE" next to each of the two statements below.
    i. A tangent line can only intersect the graph of a function once. FALSE
    ii. Any line that intersects the graph of a function exactly once *must* be the tangent line to the graph at that point. FALSE

a. Draw the graph of a function and a tangent line at $x = a$ such that the tangent line only intersects the graph once.

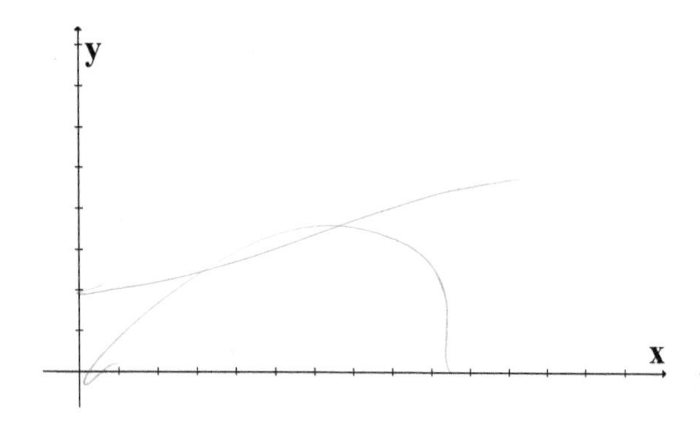

b. Draw the graph of a function and a tangent line at $x = a$ such that the tangent line intersects the graph of the function at $x = a$, but also at one or more other points.

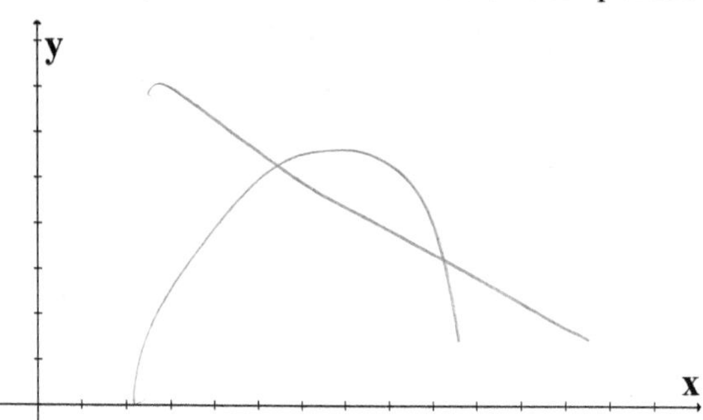

c. Draw the graph of a function and a line that is *not* a tangent line to the graph of the function at $x = a$, but intersects the graph exactly once (at $x = a$).

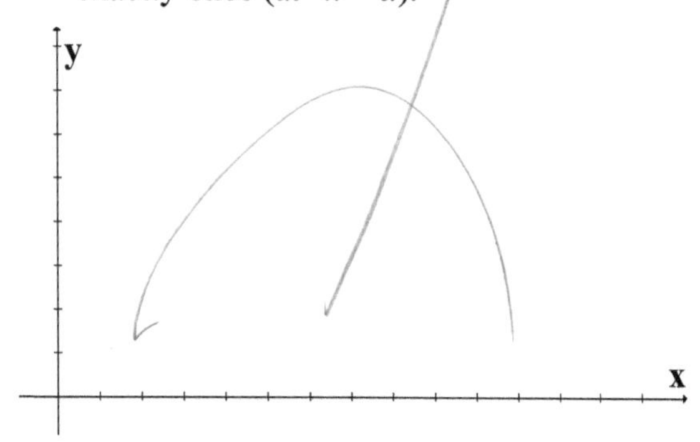

Notes

Notes

Derivatives 3: The Derivative as a Function

Model 1: Graph of a Function

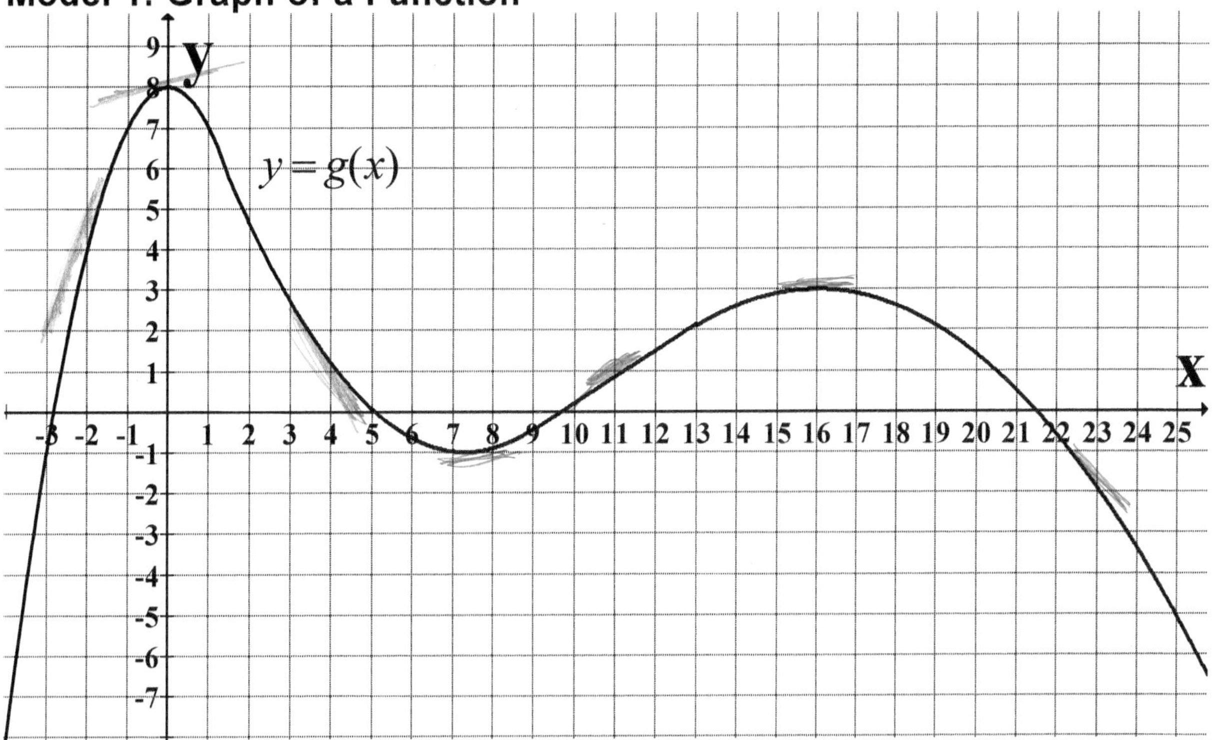

Construct Your Understanding Questions (to do in class)

1. Sketch lines tangent to the graph of g (above) at $x = -2, 0, 4, 7.3, 11, 16, 23$.

   a. Fill in the table with the slopes of these tangent lines. Choose from -4, -1.3, 0, 0.7, 1.3, or 4. Some of these values may be used more than once, and others not at all.

x	-2	0	4	7.3	11	16	23
m_{tangent}	4	0	-4	0	1.3	0	-1.3

   b. Plot each of these points (below) and sketch a smooth curve connecting the points.

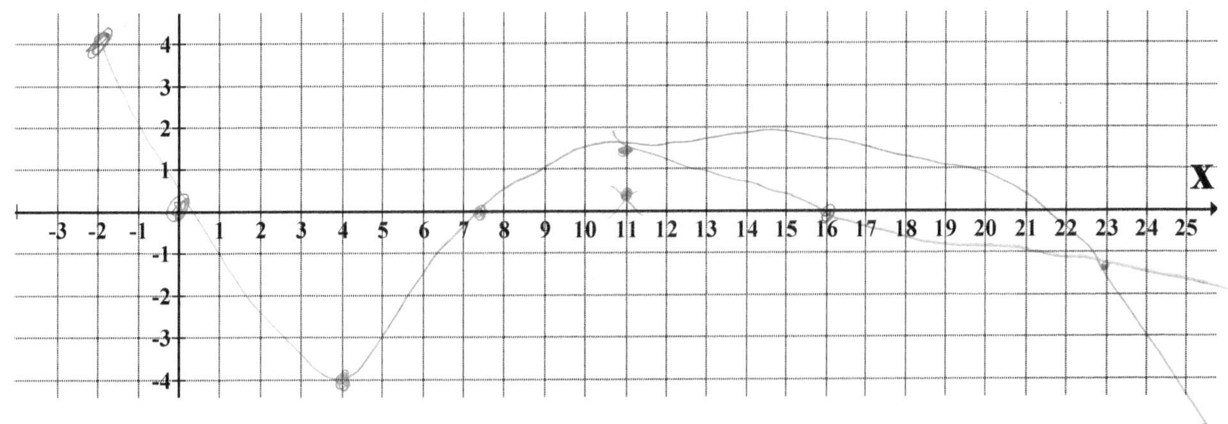

c. (Check your work) The curve you drew at the bottom of the previous page should cross the x axis three times.

2. Look at the smooth curve you drew in Question 1b at the bottom of the previous page. Consider the points on this curve between the points you plotted, for example, at $x = 2$.

   a. Use this graph to estimate the value of y associated with $x = 2$. −2

   b. What does this value of y tell you about the original function g (in Model 1)?

   Slope was negative 2

3. Estimate the slope of the line tangent to g in Model 1 at...

   a. $x = 9.5$ b. $x = 21$ Describe the method you used to make these estimates.

   1.5 −1

4. According to Summary Box D3.1, what symbol should be used for the function you drew on the graph at the bottom of the previous page? Explain your reasoning.

Summary Box D3.1: Derivative as a Function

For a function f, the function f' (as defined below) gives the slope of the line tangent to the graph of f at each point x where $f'(x)$ exists. The function f' is called the **derivative** of the function f.

$$f'(x) = \lim_{h \to 0} \frac{f(x+h) - f(x)}{h}$$

5. On the axes at right, sketch a portion of a function $k(x)$ near $x = a$ based on the following information about the derivative of the function:

   - 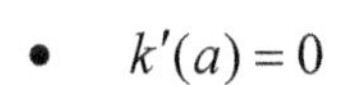 $k'(a) = 0$
   - $k'(x)$ is negative for $x < a$
   - $k'(x)$ is positive for $x > a$

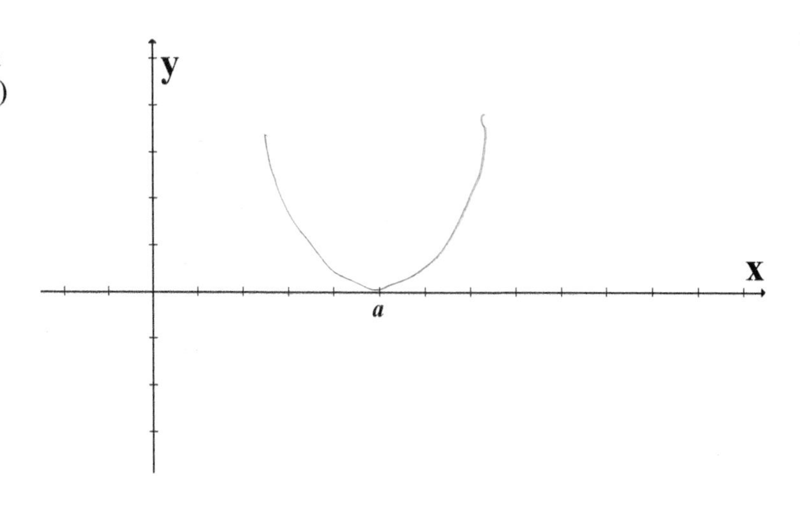

6. What can you say about a function f based on the information that f' is...

   a. zero at a given point?

   flat

   b. negative over a given interval?

   negative slope

   c. positive over a given interval?

   positive slope

7. Complete Summary Box D3.2 by filling in the blanks with **increasing** or **decreasing**, as appropriate. (Recall that increasing means that $f(x)$ is getting larger as x gets larger.)

Summary Box D3.2: Positive and Negative Derivatives

- If $f'(x) > 0$ on an interval then f is increasing on that interval.
- If $f'(x) < 0$ on an interval then f is decreasing on that interval.

8. On the axes at right, sketch a function $g(x)$ based on the information that:

   - $g'(a) = 0$
   - $g'(x) > 0$ when $x \neq a$ (that is, the derivative is positive on both sides of $x = a$)

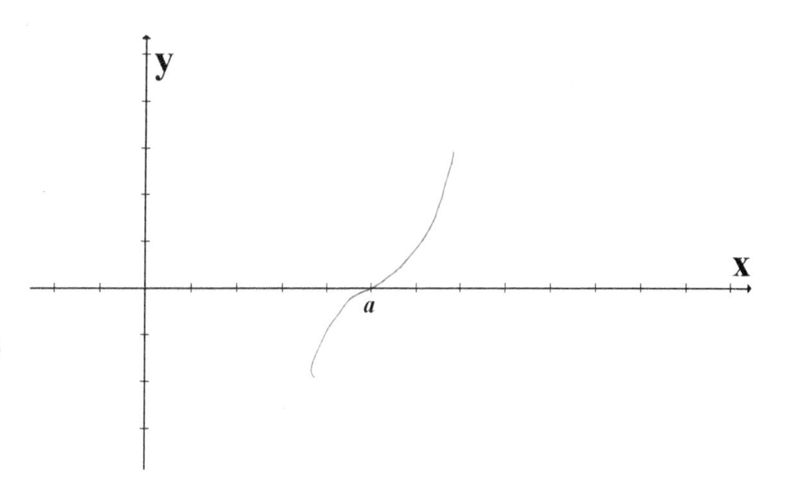

9. The graph of a function g is shown at right.
   - Mark each point where $g'(x) = 0$.
   - Label the intervals "+" if $g'(x) > 0$ or "−" if $g'(x) < 0$, as appropriate.

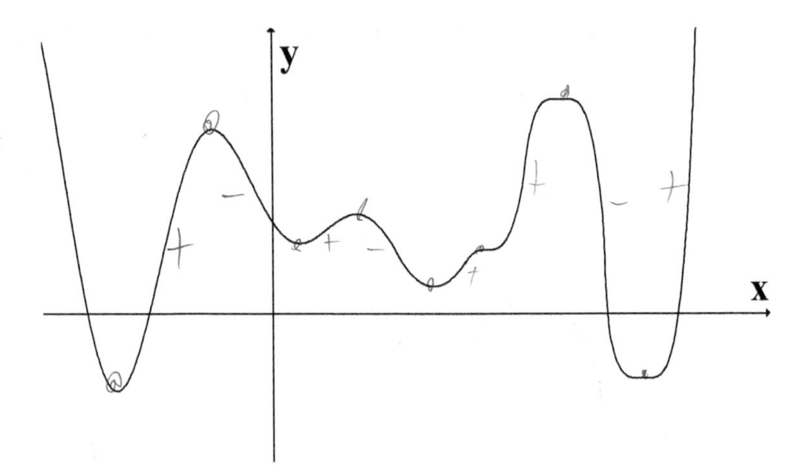

10. On **Graph iii**, sketch the derivative of f shown on **Graph i**. On **Graph iv**, sketch the derivative of f shown on **Graph ii**.

i.

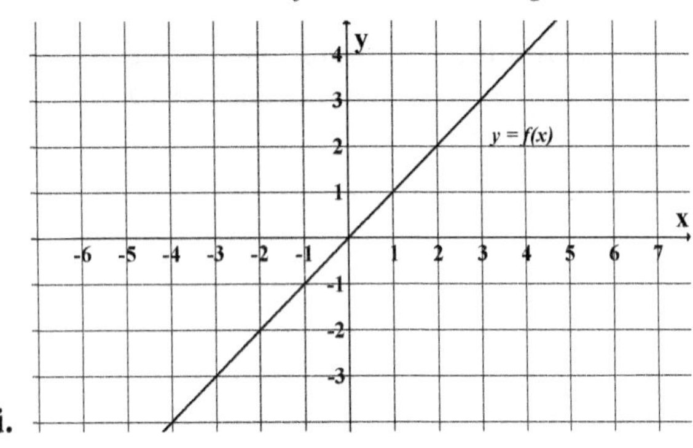

ii.

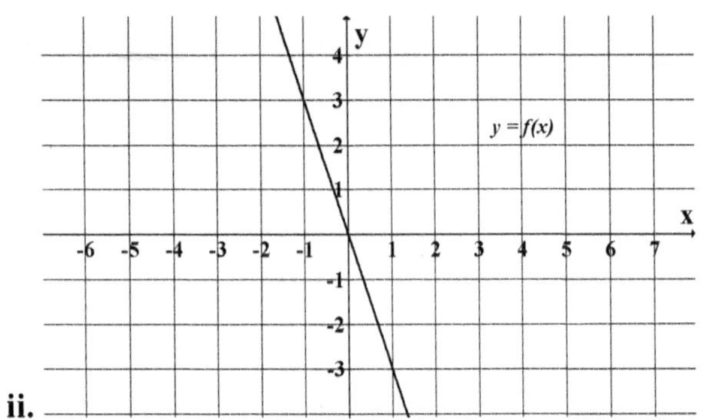

iii.

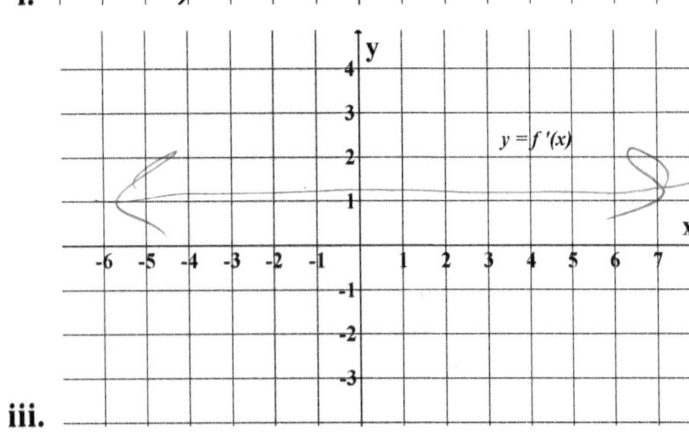

iv.

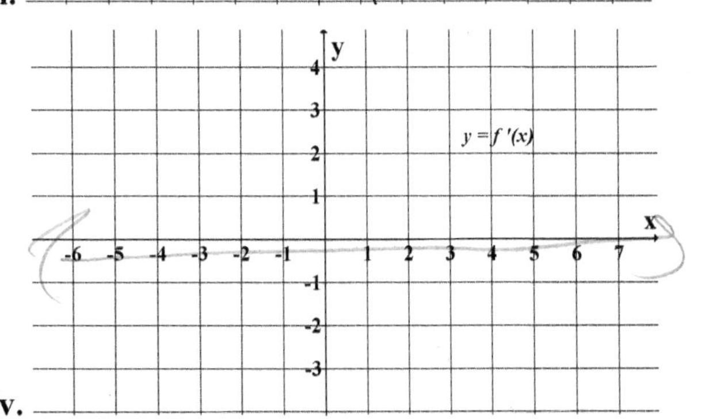

11. Circle the formula that represents the function you drew on **Graph iii**.

Put a box around the formula that represents the function you drew on **Graph iv**.

Choose from: $f'(x) = \frac{1}{3}$, $f'(x) = -\frac{1}{3}$, $f'(x) = -3x$, $f'(x) = 3x$, $f'(x) = -\frac{1}{3}x$

$f(x) = 0$, $f(x) = 1$, $f(x) = -1$, $f'(x) = -1$, $f'(x) = 1$, $f'(x) = 3$, $f'(x) = -3$,

12. (Check your work) Explain why the graph of the derivative of any linear function is a horizontal line.

no slope change

Extend Your Understanding Questions (to do in or out of class)

13. Which graph below (**vi, vii,** or **viii**) describes f' when f is the parabola shown in **Graph v**? Hint: start by finding the points where you expect the derivative of the function represented in **Graph v** to be zero or change sign (e.g., from positive to negative).

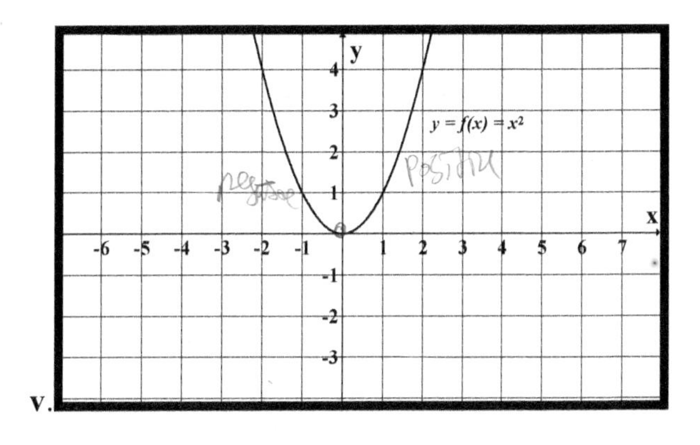

v.

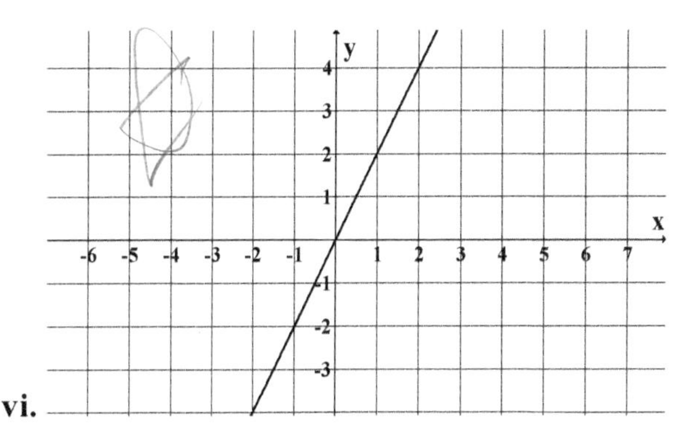

vi.

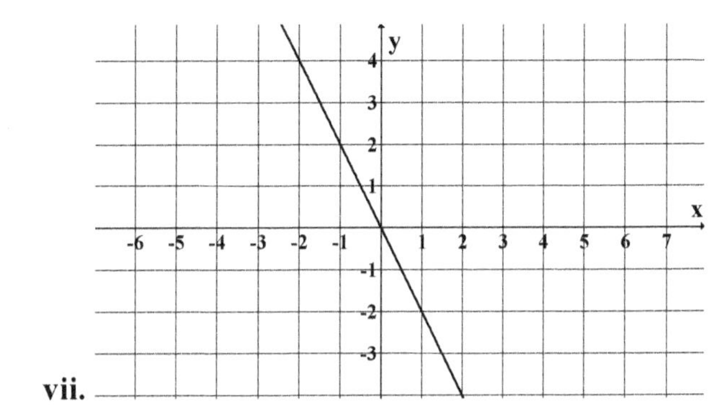

vii.

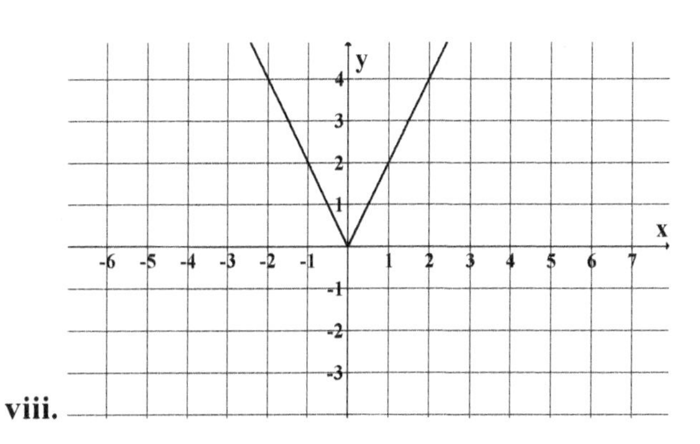

viii.

14. (Check your work) The slope of the tangent line to the parabola in **Graph v** is…

a. -2 at $x = -1$. On the graph of f' that you chose, what is $f'(-1) =$

b. 2 at $x = 1$. On the graph of f' that you chose, what is $f'(1) =$

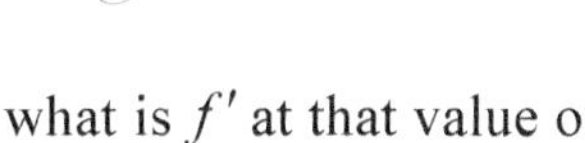

c. zero at what value of x? On the graph of f' that you chose, what is f' at that value of x?

15. **Graph vii** from the previous page (also shown below, left) is <u>not</u> the correct answer to Question 13. On the axes below, right, sketch a graph of a function f whose derivative f' is represented in **Graph vii**.

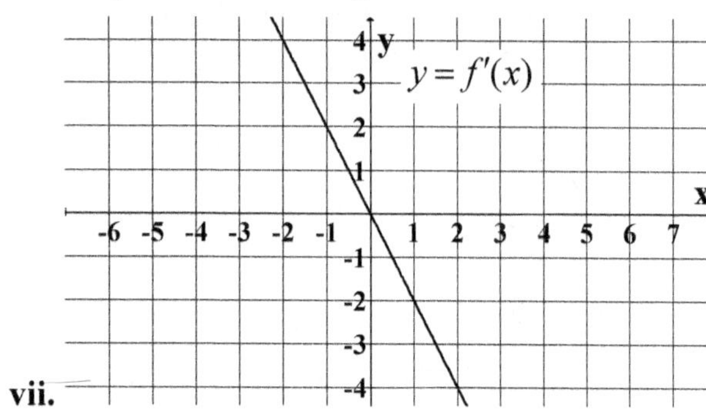

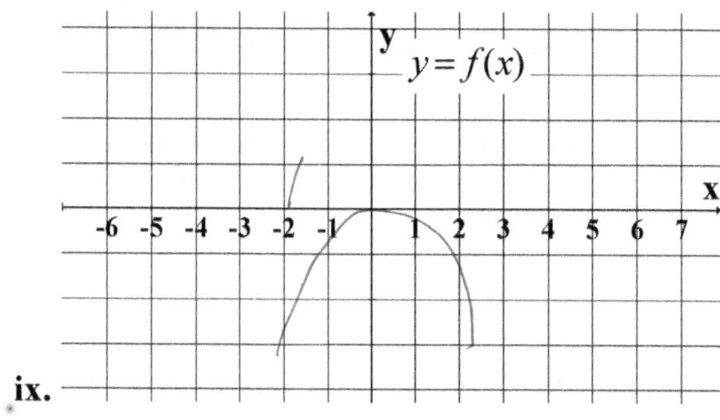

a. (Check your work) Find a point on your sketch of the graph of f where the slope of the tangent line is negative. Does this match the information on **Graph vii**? Explain.

b. (Check your work) Find a point on your sketch of the graph of f where the slope of the tangent line is positive. Does this match the information on **Graph vii**? Explain.

16. (Check your work) The two graphs below are <u>both</u> correct answers to the previous question.

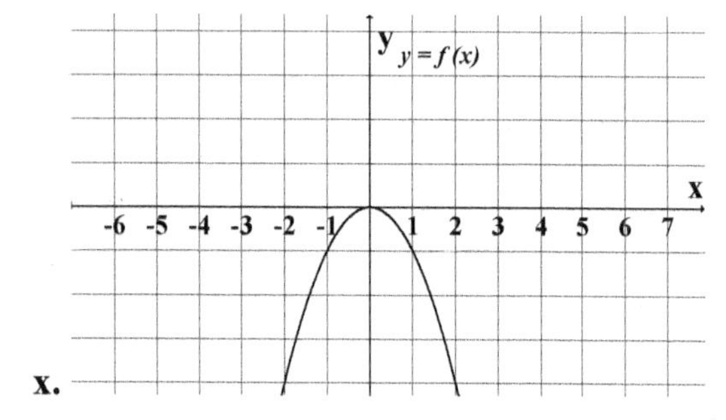

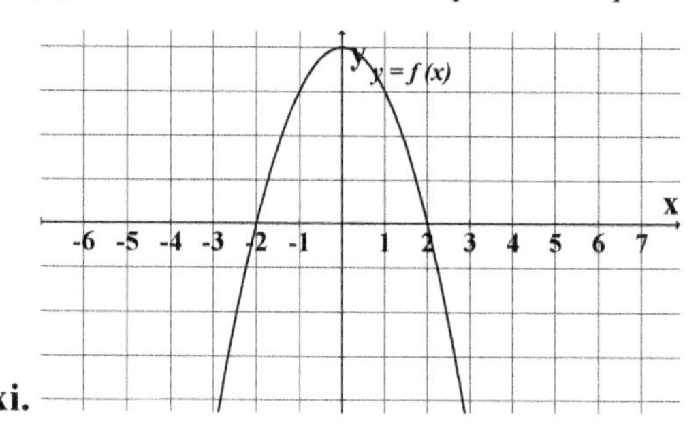

a. Construct an explanation for why both answers are acceptable.

b. It turns out there are an infinite number of correct answers to Question 15. Sketch several of these, and explain why shifting the parabola up or down does not change the graph of f' (**Graph vii**).

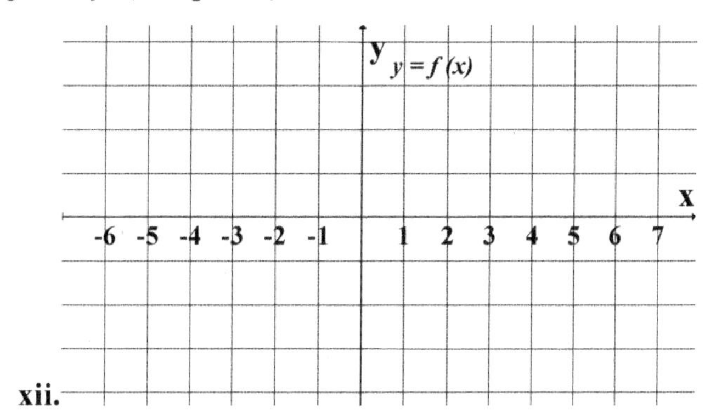

17. Assume f is increasing over an interval. Select all that are TRUE.

   a. f' is positive on that interval
   b. f' is increasing on that interval
   c. the graph of f' is below the $x-$axis on that interval
   d. f' is negative on that interval
   e. the graph of f' is above the $x-$axis on that interval
   f. f' is decreasing on that interval
   g. f is positive on that interval

18. (Check your work) Only two statements in the previous question are true (and statement b is not one of them). Explain why statement b. is false, and support your explanation by drawing an example of a function f that is increasing while its derivative, f', is decreasing.

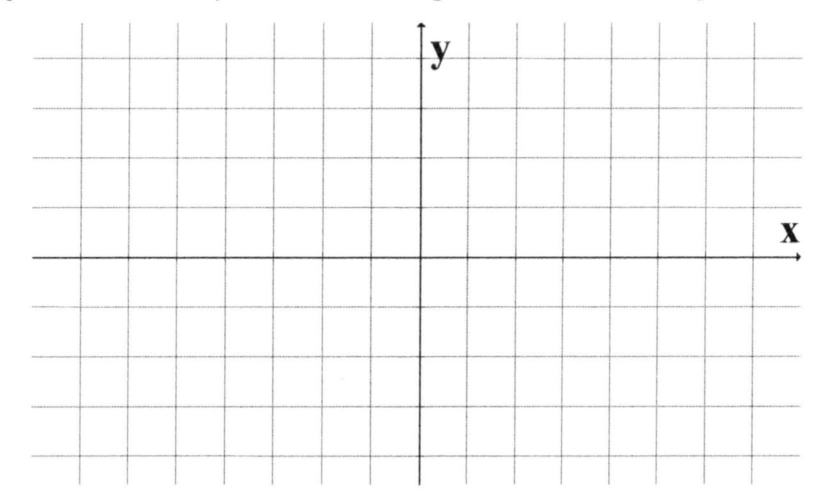

Notes

Derivatives 4: Differentiability

Model 1: Approaching a Point from the Left or Right

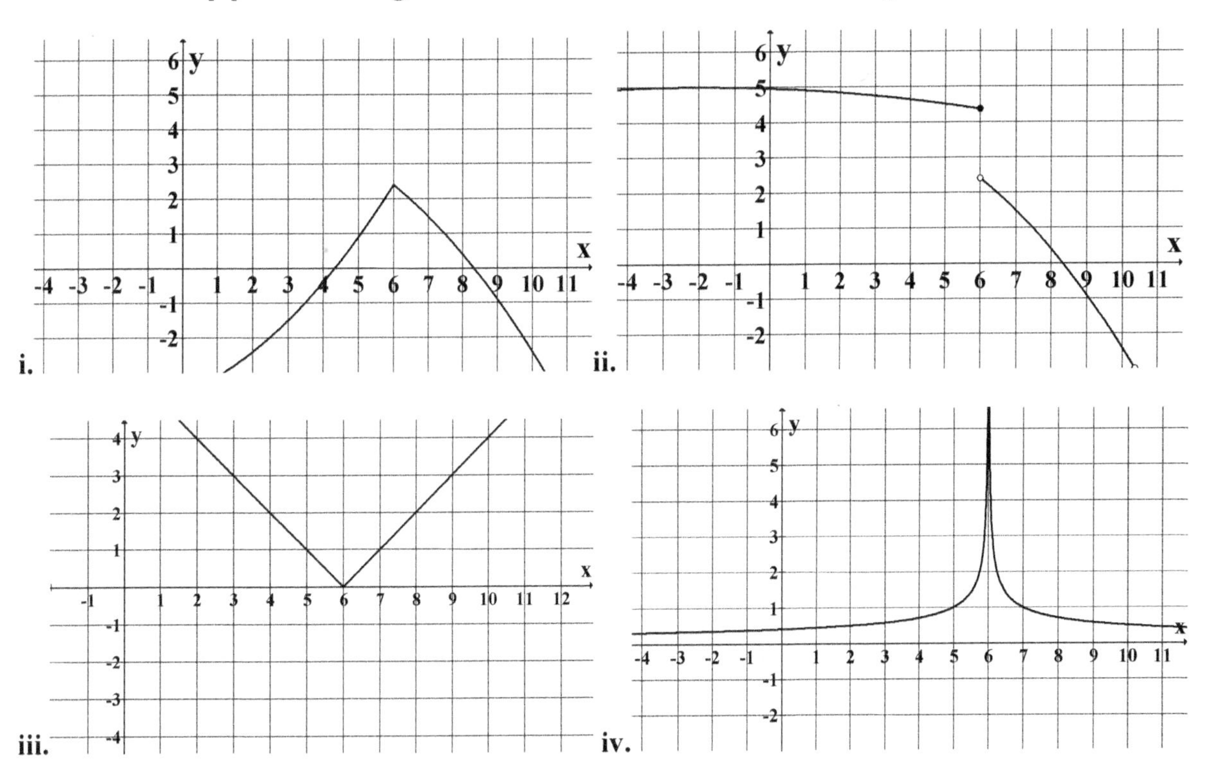

Construct Your Understanding Questions (to do in class)

1. What does the derivative at a point (if it exists) tell you about the function at that point?

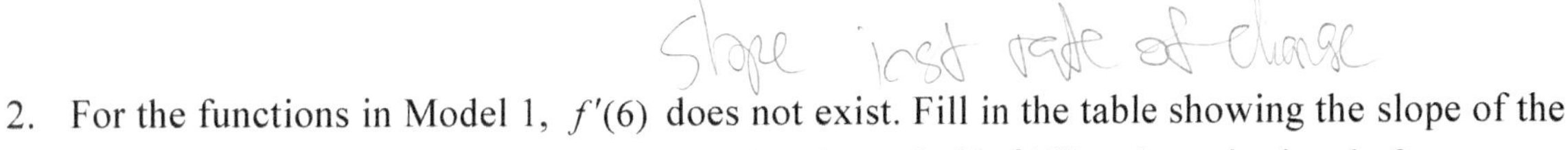

2. For the functions in Model 1, $f'(6)$ does not exist. Fill in the table showing the slope of the secant line to each curve in Model 1 passing through $(6, f(6))$ and a point just before: $(5.9, f(5.9))$, or just after: $(6.1, f(6.1))$. The first two are done for you. For **Graph iv**, do not give a value. Instead, describe the slope in words, including whether it is positive or negative.

	Slope of the line through $(5.9, f(5.9))$ and $(6, f(6))$	Slope of the line through $(6, f(6))$ and $(6.1, f(6.1))$
Graph i	1.6	-0.82
Graph ii	-0.17	-0.82
Graph iii	-1	1
Graph iv		

3. Construct an explanation for why the derivative of each function in Model 1 does not exist at $x = 6$.

4. (Check your work) Is your answer to the previous question consistent with Summary Box D4.1? Explain you reasoning. Write down any questions that you have for the instructor.

Summary Box D4.1: Differentiable and Non-differentiable

Recall that for a limit to exist at a point a, it must be the same as you approach a from the left and the right. Since the derivative at a is defined as the limit of the difference quotient shown below, $f'(a)$ exists if and only if...

$$\lim_{h\to 0^+}\frac{f(a+h)-f(a)}{h}=\lim_{h\to 0^-}\frac{f(a+h)-f(a)}{h}=\text{real number}$$

If $f'(a)$ exists then f is **differentiable** at a. If it does not exist, then f is **non-differentiable** at a.

5. **True** or **False**: According to Summary Box D4.1, if the graph of a function f has a vertical tangent line at a, then $f'(a)$ does not exist. Circle one and explain your reasoning.

Model 2a: Removable (Point) Discontinuity

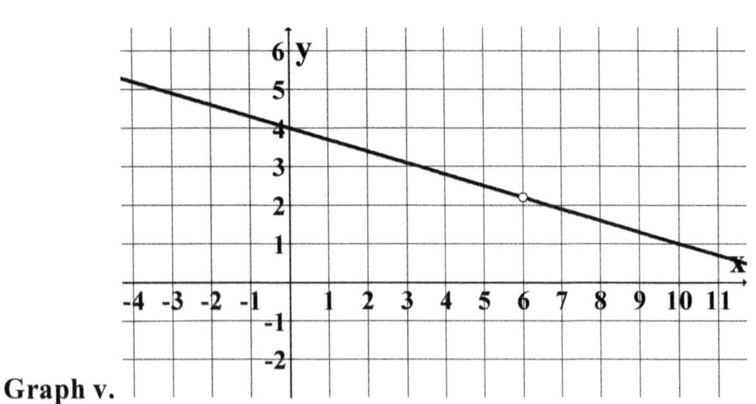

Graph v.

Construct Your Understanding Questions (to do in class)

6. For **Graph v** in Model 2a...
   a. Does $f(6)$ exist? yes
   b. Circle each of the following that accurately defines the derivative: $f'(6)=$

   $\lim_{h\to 0}\frac{f(6+h)-f(6)}{h}$ $\qquad$ $\lim_{x\to 6}\frac{f(x)-f(6)}{x-6}$ $\qquad$ $\lim_{x\to 6} f(x)$

   c. Decide if $f'(6)$ exists and explain your reasoning.

7. Adam cites the following arguments as evidence that $f'(6)$ exists for **Graph v**.

   - $\lim_{x\to 6^-} f(x) = \lim_{x\to 6^+} f(x) =$ real number.
   - The slope of the tangent line to the graph of f immediately before $x = 6$ is equal to the slope of the tangent line to f immediately after $x = 6$.

   a. Mark each bulleted statement above **TRUE** or **FALSE** for f in **Graph v**.

   b. (Check your work) Many students conclude that because both bulleted statements are true, the function f on **Graph v** must be differentiable at $x = 6$. It is not. If you used one of these arguments to claim $f'(6)$ exists for **Graph v**, go back and reconsider your answer to part c of the previous question.

Model 2b: Removable Discontinuity (where $f(a)$ exists)

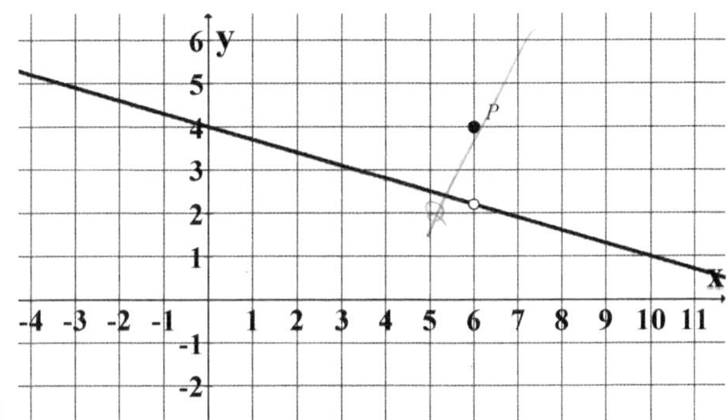

Graph vi.

Construct Your Understanding Questions (to do in class)

8. For **Graph vi** in Model 2b...

   a. What is $f(6) =$ 4

   b. Mark a point Q on the graph whose x value is close to 6, but not equal to 6.

   c. Draw the secant line through this point Q and the point marked P on the graph.

   d. Describe what happens to the slope of this secant line as Q moves closer to P.

   e. What does your answer to part d tell you about the differentiability of f on **Graph vi** at $x = 6$? (Is the function in **Graph vi** differentiable at $x = 6$?)

9. Specify each value of x where the function represented in the graph is <u>not</u> differentiable.

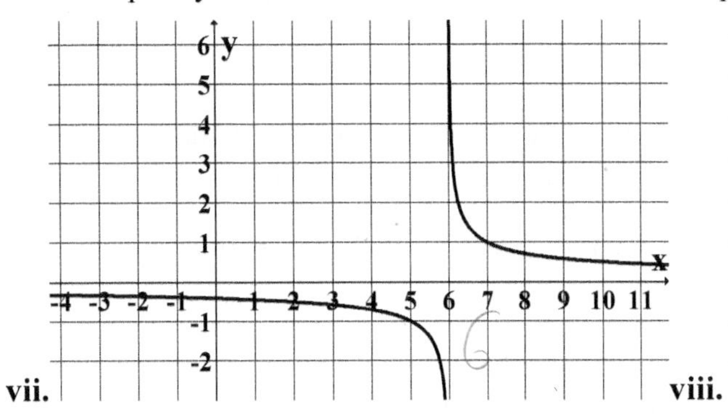

vii.

viii.

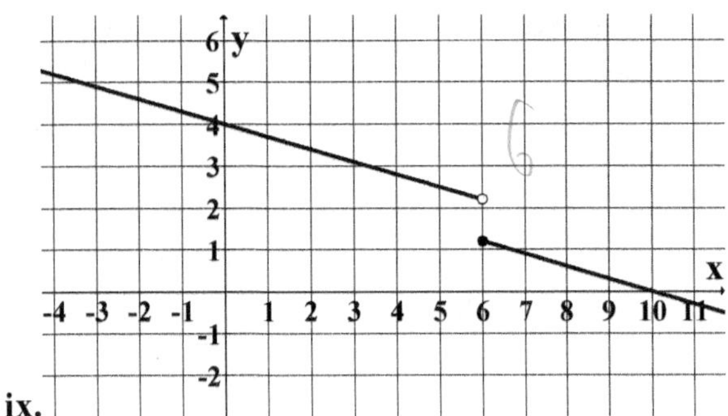

ix.

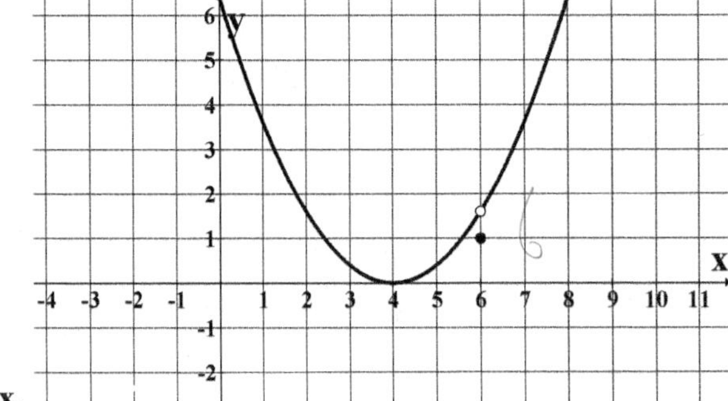

x.

xi.

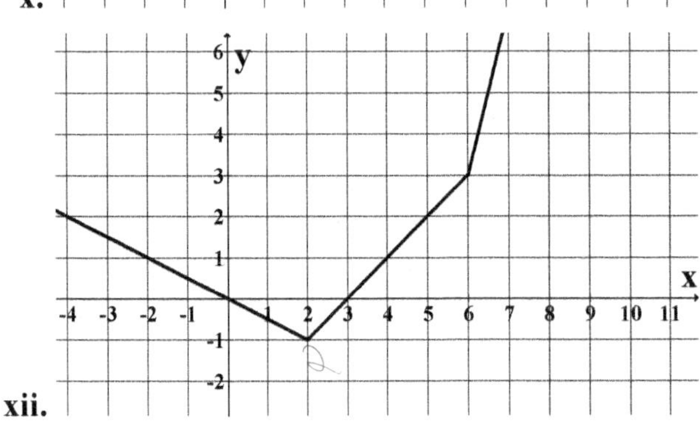

xii.

10. Based on Summary Box D4.2 (on the next page), mark each statement **True** <u>or</u> **False**.
<u>For each statement that is false</u>, sketch a function that demonstrates the statement is false.

a. If f is continuous over an interval then f is differentiable over that interval.

b. If f is differentiable over an interval then f is continuous over that interval.

c. If f has a discontinuity at a then $f'(a)$ does not exist.

Summary Box D4.2

A function must be continuous at $x = a$ to be differentiable at $x = a$.

Extend Your Understanding Questions (to do in class)

11. Only one of the following functions is differentiable for all x. Identify this function. For the others, identify the value of x where the function is not differentiable.

a. $f(x) = \dfrac{x^2 - 4}{x + 2}$

b. $f(x) = \dfrac{1}{1 - x}$

c. $f(x) = x^4 - 3x^2 + 2x + 1$

d. $f(x) = \begin{cases} 2x, & x < 2 \\ 4, & x \geq 2 \end{cases}$

e. $f(x) = 3 - 2|x - 6|$

12. (Check your work) Statements A-E, below, are all TRUE.

For each graph in this activity that represents a function that is not differentiable at a point, go back and label the point with the letter of the one statement that best explains why that function is NOT differentiable at that point. (Some statements may be used more than once; others, not at all.)

A. $f(a)$ appears in the definition of the derivative $\left(f'(a) = \lim_{h \to 0} \dfrac{f(a+h) - f(a)}{h} \right)$, so if $f(a)$ does not exist then $f'(a)$ does not exist.

B. $f'(a)$ does not exist if there is a vertical tangent line to the graph of $f(x)$ at $x = a$.

C. $f'(a)$ does not exist when $\lim_{x \to a} f(x) \neq f(a)$.

D. $f'(a)$ does not exist when $\lim_{x \to a} f(x)$ does not exist.

E. $f'(a)$ does not exist if $\lim_{h \to 0^+} \dfrac{f(a+h) - f(a)}{h} \neq \lim_{h \to 0^-} \dfrac{f(a+h) - f(a)}{h}$.

Notes

Derivatives 5: The Second Derivative

Model 1: Derivatives Revisited

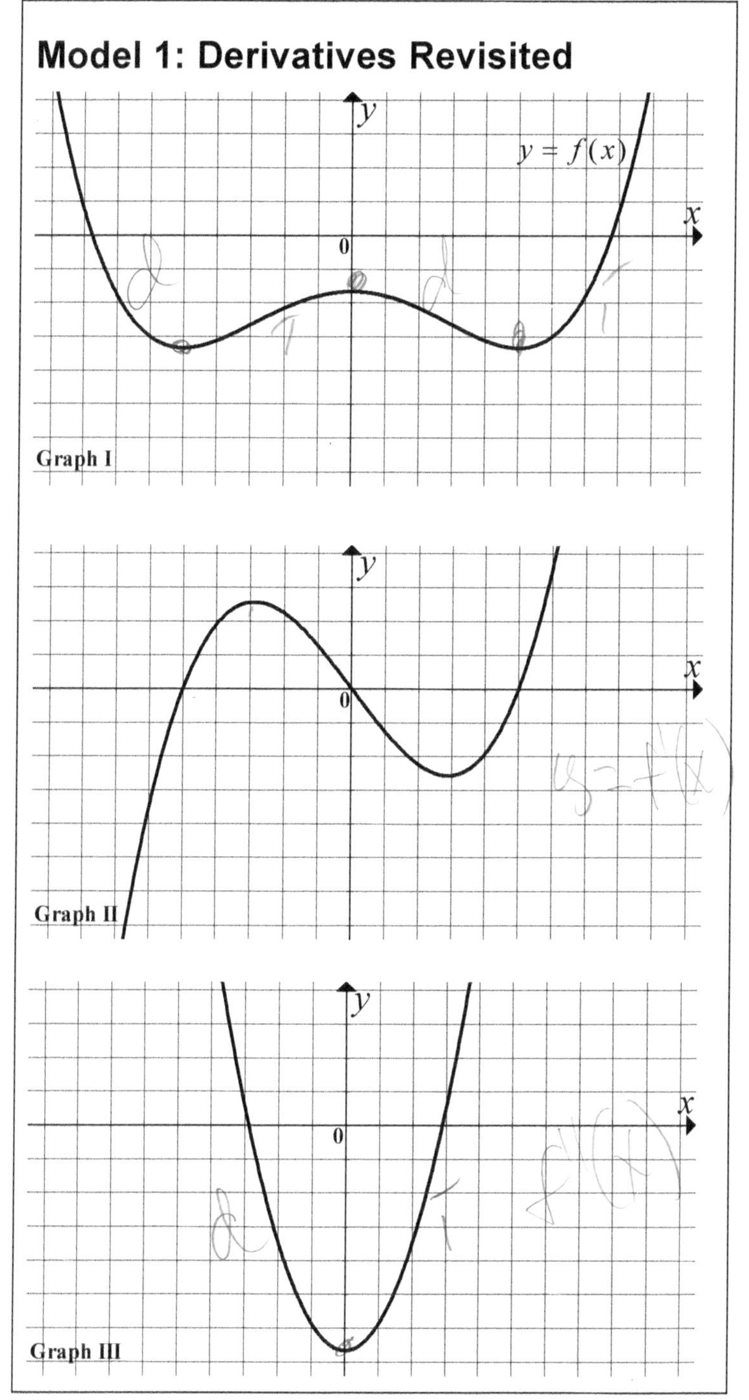

Construct Your Understanding

1. For f on **Graph I**...
   a. mark each point where $f'(x) = 0$.
   b. mark each interval along the $x-$ axis where $f'(x) < 0$ with a "d".
   c. mark each interval along the $x-$ axis where $f'(x) > 0$ with a "i".
2. What do the "i" and "d" in the previous question stand for?
3. **True** or **False**: **Graph II** could be the derivative of **Graph I**. If false, mark a point where **Graph II** does not seem to equal to $f'(a)$.
4. Answer Question 1 for **Graph II**. Then decide if **Graph III** could be the derivative of **Graph II**. Explain your reasoning.

Summary Box D5.1: The Second Derivative

The second derivative of a function $y = f(x)$ can be represented as ...

$$f'',\ y'',\ \frac{d}{dx}\left(\frac{dy}{dx}\right),\ \frac{d^2y}{dx^2},\ \text{or}\ f''(x).$$

5. (Check your work) Mark **Graph II** with $y = f'(x)$, and **Graph III** with $y = f''(x)$. Find the values of x where $f''(x) = 0$, then describe what is happening to the graph of...
   a. f (**Graph I**) at these values of x.
   b. f' (**Graph II**) at these values of x.

6. Mark the intervals where $f''(x) > 0$ and $f''(x) < 0$, respectively. Describe what is happening to the graph of f (**Graph I**) when…

   a. $f''(x) > 0$ b. $f''(x) < 0$

7. Describe what is happening to the graph of f' (**Graph II**) when…

   a. $f''(x) > 0$ b. $f''(x) < 0$

8. A student examines the portion of the function g shown at right, and says that g' is decreasing on the interval from $x = 0$ to $x = a$. He gives the following reasoning: "If the graph were a ski slope, it would be getting less steep from left to right, so the derivative must be decreasing from left to right."

   Do you **agree** or **disagree** with his statement and reasoning? Circle one and explain your reasoning.

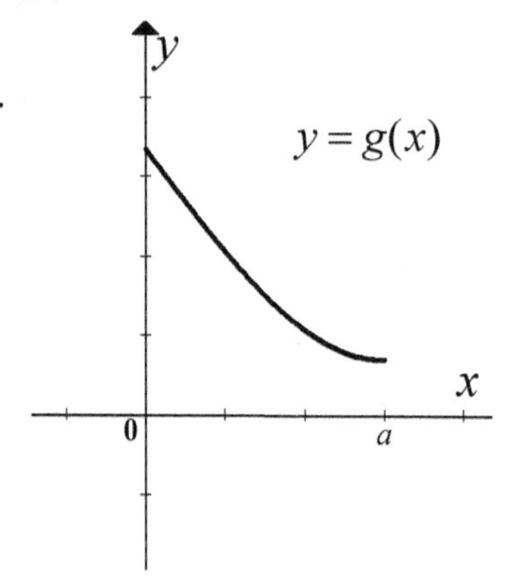

9. (Check your work) Find a portion of a graph in Model 1 that mimics the curvature of g in the previous question, then check if the derivative (shown in the graph below it) is increasing or decreasing over that interval. Is this consistent with your answer to the previous question? If not, you should also check your answer to Question 7.

10. Fill in the blanks. Choose from: **concave up, concave down, increasing, decreasing, positive, negative, zero.** (Some terms will be used more than once, others not at all.)

    a. When f' is increasing, f'' is ________________, and f is ________________.

    b. When f' is decreasing, f'' is ________________, and f is ________________.

    c. When f is increasing, f' is ________________.

    d. When f is decreasing, f' is ________________.

11. Complete Summary Box D5.2, by filling in each blank. Choose from the following terms: **concave up, concave down, increasing, decreasing, positive, negative, zero**. Some terms may be used more than once.

Summary Box D5.2: Interpretation of the First and Second Derivatives

If $f'(x) > 0$ on an interval, then f is ______________________________ on that interval.

If $f'(x) < 0$ on an interval, then f is ______________________________ on that interval.

If $f'(x)$ is __________________ on an interval, then f is neither ______________________________,

nor ____________________________ on that interval.

If $f''(x) > 0$ on an interval, then f is ______________________________ on that interval.

If $f''(x) < 0$ on an interval, then f is ______________________________ on that interval.

If $f''(x)$ changes from ________________________ to ________________________ (or vice versa) at a point P on the graph of f, then P is an **inflection point**.

If $f''(x) > 0$ on an interval, then f' is ______________________________ on that interval.

If $f''(x) < 0$ on an interval, then f' is ______________________________ on that interval.

12. For each graph below, estimate and mark the following:
    a. On the graph, mark all points where $f'(a) = 0$.
    b. On the x-axis, mark each interval where $f'(x) < 0$, and $f'(x) > 0$. (Mark these **i**, or **d**, as appropriate.)
    c. On the graph, mark all inflection points (**IP**).
    d. On the x-axis, mark each interval where $f''(x) < 0$; and $f''(x) > 0$. (Mark these **CD** or **CU**, as appropriate.)

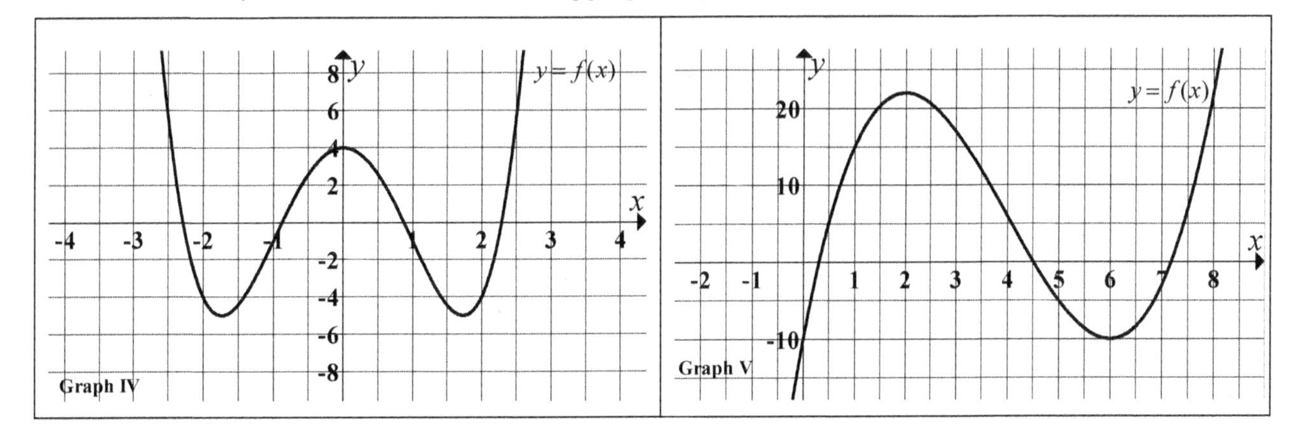

13. Mark each inflection point on the graphs below.

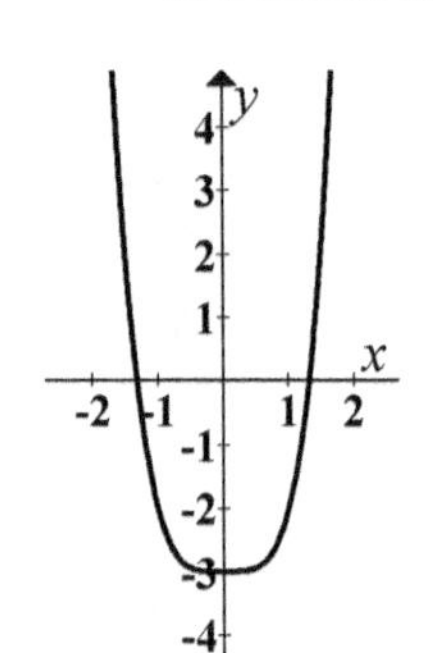

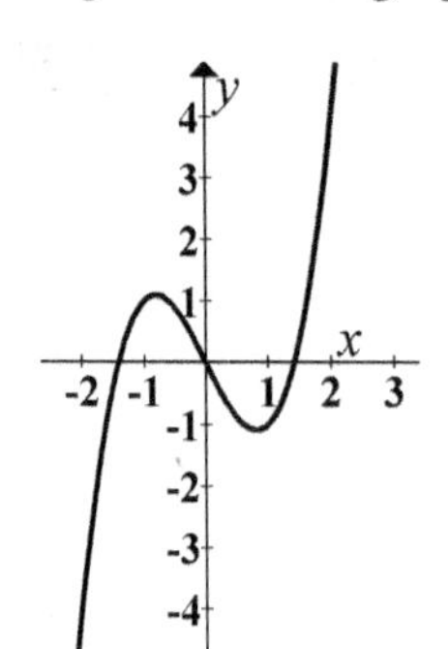

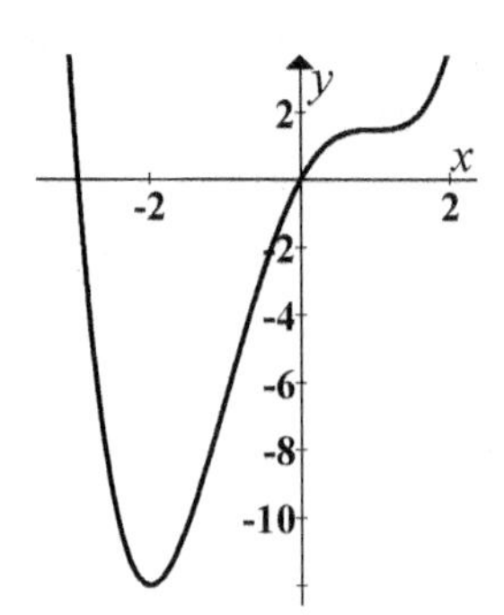

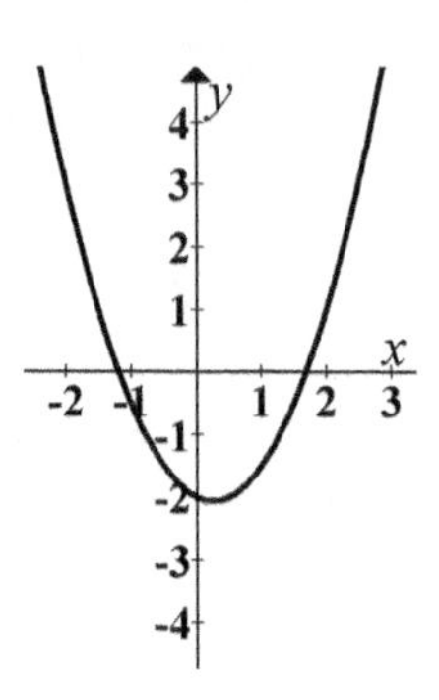

14. Given the following equations for the functions f graphed in the previous question, find f'', then find each point where $f''(x) = 0$. Show your work.

$f(x) = x^4 - 3$	$f(x) = x^3 - 2x$	$f(x) = \frac{1}{2}x^4 - 3x^2 + 4x$	$f(x) = x^2 - \frac{1}{2}x - 2$

15. Circle True or False, and cite evidence from this activity to support your choice.

a. **True** or **False:** If $f''(a) = 0$, then there is an inflection point at $x = a$.

b. **True** or **False:** If there is an inflection point at $x = a$ then $f''(a) = 0$.

16. (Check your work) The following is a true statement:

f'' must be zero or undefined at $x = a$ if f has an inflection point at $x = a$.

Is this consistent with your answers above and the definition of an inflection point in Summary Box D5.2? Explain.

Notes

Notes

Differentiation Techniques 1: Power, Constant Multiple, Sum and Difference Rules

Model 1: Finding the Equation of $f'(x)$ from a Graph of $f(x)$

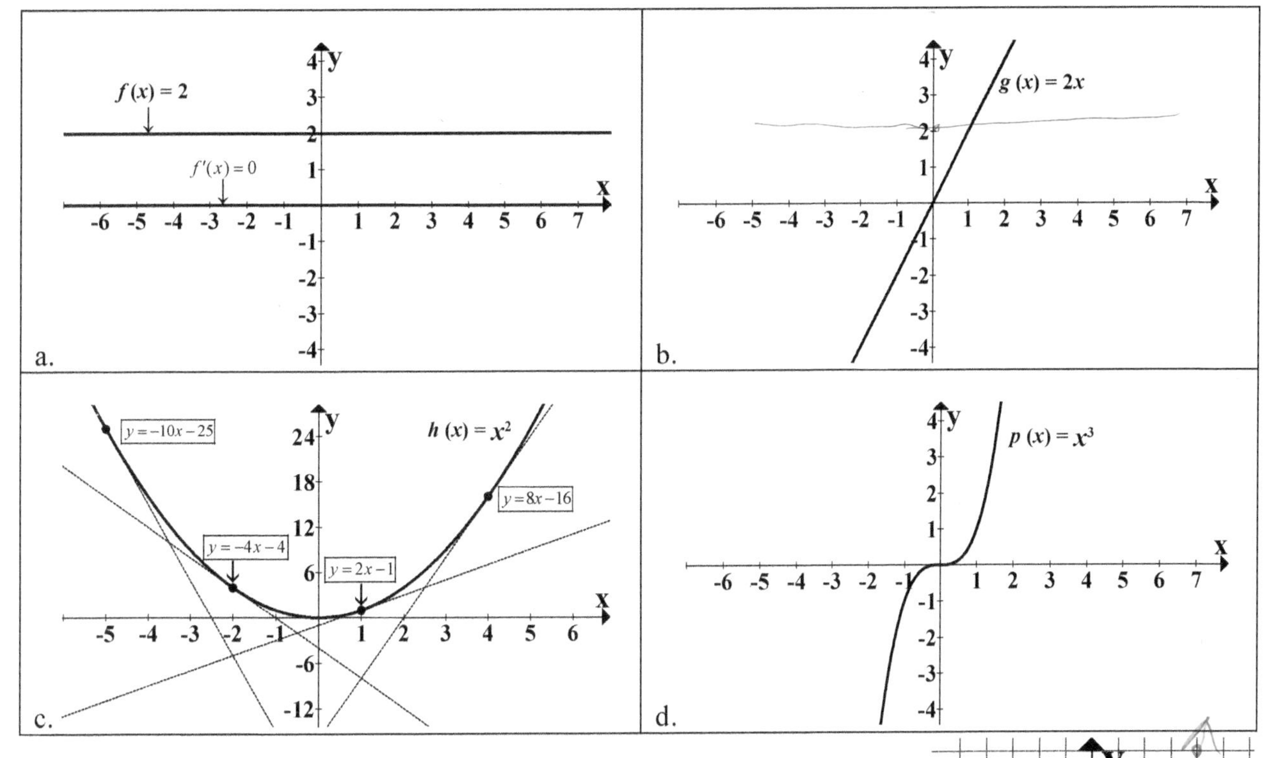

Construct Your Understanding Questions

1. Consider **Graph a** in Model 1.

   a. What is the slope of the line $f(x) = 2$?

   b. Is your answer consistent with the graph of its derivative, $f'(x) = 0$?

2. Now consider **Graph b** in Model 1.

   a. On **Graph b**, sketch the derivative of the function $g(x) = 2x$, shown there.

   b. Determine the equation of this derivative: $g'(x) =$

3. Now consider **Graph c** in Model 1.

   a. On the axes (above, right) sketch the derivative of the function shown in **Graph c**. To help, the equations of tangent lines to the graph at $x = -5, -2, 1$, and 4 are given.

   b. From your graph, determine the equation of this derivative: $h'(x) =$

   c. f in **Graph a** is the derivative of the function g in **Graph b**. Does the same relationship hold for g and h in **Graph b** and **Graph c**?

4. (Check your work) f (on **Graph a**) is the derivative of g (on **Graph b**) and g is the derivative of h (on **Graph c**), so Kai guesses that h must be the derivative of p on **Graph d**. To check if this is reasonable, he makes the following table. Complete his table by filling in empty boxes with **positive**, **negative**, or **zero**.

x	Slope of line tangent to p in **Graph d**	Value of the function h in **Graph c**
$x=-1$	positive	positive
$x=0$	zero	
$x=1$		

a. Kai concludes that h on **Graph c** is a reasonable guess for the derivative of p on **Graph d**. Explain how the data in the table above supports this conclusion.

b. The instructor asks Kai to explain his reasoning. He begins…

"x^2 is always positive except at $x = 0$, where it is zero, and

the slopes of lines tangent to $f(x) = x^3$ are positive except at $x = 0$, where it is zero."

Mark each underlined statement TRUE or FALSE.

c. (Check your work) The instructor says: "I like your thinking Kai, but lines tangent to any curve of the form $f(x) = k \cdot x^3$ (where k is a positive constant) will be zero or positive." Is this consistent with your answer to part b?

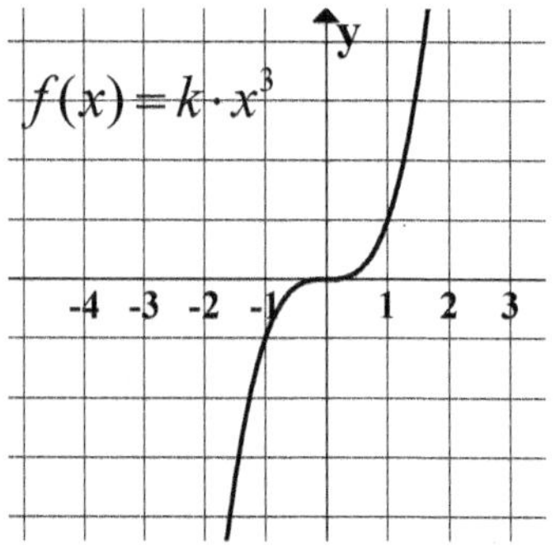

d. To figure out if Kai's guess is correct, the instructor suggests he use the definition of the derivative

$$\frac{d}{dx}(f(x)) = \lim_{h\to 0}\frac{f(x+h)-f(x)}{h}$$

to find an expression for $\frac{d}{dx}(x^3)$.

A partial solution is shown on the next page, but take a moment to derive an expression for $\frac{d}{dx}(x^3)$ without looking ahead.

e. (Check your work) Is Kai's work shown at right consistent with your work on the previous page?

f. Is $\frac{d}{dx}\left(x^3\right) = x^2$, as Kai was expecting? If not, what is $\frac{d}{dx}\left(x^3\right) =$

Kai's Work on $\frac{d}{dx}\left(x^3\right)$
$\frac{d}{dx}\left(x^3\right) = \lim_{h\to 0}\frac{(x+h)^3 - x^3}{h}$ $\frac{d}{dx}\left(x^3\right) = \lim_{h\to 0}\frac{x^3 + 3x^2h + 3xh^2 + h^3 - x^3}{h}$ $\frac{d}{dx}\left(x^3\right) = \lim_{h\to 0}\frac{3x^2h + 3xh^2 + h^3}{h}$ $\frac{d}{dx}\left(x^3\right) = \lim_{h\to 0}\frac{h(3x^2 + 3xh + h^2)}{h}$

5. For functions of the form $f(x) = x^n$, we call "n" the **power of x**. Describe in words how the power of x changed in all the examples so far in this activity when going from a function f to its derivative f'.

6. (Check your work) Look at the entries in the table in Model 2a (found on the next page), and decide if your answer to the previous question is consistent. If not, reconsider the previous question.

Model 2a: Derivatives of Power Functions

For a Function, $f(x)$...		and its Derivative, $f'(x)$	
Equation of $f(x)$	**Power of x**	**Equation of $f'(x)$**	**Power of x**
$f(x) = x^2$	2	$f'(x) = 2x$	1
$f(x) = x^3$	3	$f'(x) = 3x^2$	2
$f(x) = x^4$	4	$f'(x) = 4x^3$	3
$f(x) = x^5$	5	$f'(x) = 5x^4$	4
$f(x) = x^{10}$	10	$f'(x) = 10x^9$	9
$f(x) = x$	1	$f'(x) = 1$	
$f(x) = x^{-1}$	−1	$f'(x) = -x^{-2}$	
$f(x) = x^{\frac{1}{2}}$	$\frac{1}{2}$	$f'(x) = \frac{1}{2}x^{-\frac{1}{2}}$	
$f(x) = x^{-\frac{1}{3}}$	$\frac{1}{3}$	$f'(x) =$	
$f(x) = x^{100}$	100	$f'(x) =$	
$f(x) = x^{-2}$	−2	$f'(x) =$	

Construct Your Understanding Questions (to do in class)

7. Complete the first five rows of the table in Model 2a. Some entries are done for you.
8. Based on the information in the first five rows of Model 2a...
   a. Write a rule describing how to find the derivative of a function $f(x) = x^n$, where n is a constant.

   b. Use this rule to complete the shaded (bottom six) rows of the table in Model 2a.

   c. (Check your work) The derivative in the last row of Model 2a should be $f'(x) = \frac{-2}{x^3}$. If this does not fit, go back and reconsider parts a and b of this question.

9. (Check your work) Are your answers to the previous two questions consistent with Summary Box DT1.1? If not, go back and revise your answers.

Summary Box DT1.1: Power Rule

$$\frac{d}{dx}\left(x^n\right) = nx^{n-1} \quad \text{where } n \text{ is a real number}$$

10. Each function in the table below can be written in the form $f(x) = x^n$. Complete the table.

$f(x)$	n	$f(x)$ written in the form x^n	$f'(x)$
$f(x) = \frac{1}{x}$			
$f(x) = \sqrt{x}$			
$f(x) = 1$			
$f(x) = \frac{1}{\sqrt[3]{x}}$			
$f(x) = x^2 \cdot x^3$			

11. (Check your work) All but one function in the previous question appears in the first column of Model 2a. Are your answers above consistent with your answers in Model 2a?

Model 2b: Derivatives of Power Functions

For functions of the form $f(x) = cx^n$, we call "n" the **power of x** and "c" the **coefficient of** x^n.

For a Function, $f(x)$...			and its Derivative, $f'(x)$		
Equation of $f(x)$	Power of x	Coefficient of x	Equation of $f'(x)$	Power of x	Coefficient of x
$f(x) = 12x$	1	12	$f'(x) = 12$	0	12
$f(x) = 10x^2$	2	10	$f'(x) = 20x$	1	20
$f(x) = 3x^2$	2		$f'(x) = 6x$	1	
$f(x) = 11x^3$	3		$f'(x) = 33x^2$	2	
$f(x) = -x^5$	5		$f'(x) = -5x^4$	4	
$f(x) = 7x^3$	3		$f'(x) =$		
$f(x) = 8x^{-1}$	-1		$f'(x) =$		
$f(x) = 2x^{\frac{1}{2}}$	$\frac{1}{2}$		$f'(x) =$		

Note: Questions exploring Model 3 are found on the next page.

Construct Your Understanding Questions (to do in class)

12. Complete the table in Model 2b. Some entries are done for you.

13. For a function $f(x) = cx^n$, write an expression in terms of n, c, and x for $f'(x) =$

Model 3: Derivative Rules from Limit Laws

Limit Law (Review from CA L2) If the limit as x approaches a of $F(x)$ and limit as x approaches a of $G(x)$ exist, then…	**Corresponding Derivative Rule** If f and g are differentiable then…
$\lim_{x\to a}[cF(x)] = c\lim_{x\to a}F(x)$ **(Limit of a Constant Times a Function Law)**	$\frac{d}{dx}[cf(x)] = cf'(x)$ **(Constant Multiple Rule)**
$\lim_{x\to a}[F(x)+G(x)] = \lim_{x\to a}F(x) + \lim_{x\to a}G(x)$ **(Limit of a Sum of Functions Law)**	$\frac{d}{dx}[f(x)+g(x)] =$ **(Sum Rule)**
$\lim_{x\to a}[F(x)-G(x)] = \lim_{x\to a}F(x) - \lim_{x\to a}G(x)$ **(Limit of a Difference of Functions Law)**	$\frac{d}{dx}[f(x)-g(x)] =$ **(Difference Rule)**

Construct Your Understanding Questions

14. a. Which Derivative Rule is proved at right?

    b. Which Limit Law is used in the proof? Mark this step.

Proof $\frac{d}{dx}[cf(x)] = \lim_{h\to 0}\frac{cf(x+h)-cf(x)}{h}$

$$= \lim_{h\to 0} c\left[\frac{f(x+h)-f(x)}{h}\right]$$

$$= c\lim_{h\to 0}\frac{f(x+h)-f(x)}{h} = cf'(x)$$

15. Complete the shaded boxes on the table in Model 3.

Extend Your Understanding Questions (to do in or out of class)

16. Find the derivative of each function.

a. $f(x) = x^4 + 2x$

b. $g(x) = 3x^6 - x^4$

c. $f(x) = -x^4 + x^3 + 3x$

d. $f(x) = 2x^4 - x^3 + 12x^2 - 15x + 1$

e. $h(x) = x + \frac{1}{x^2} + 9$

f. $f(t) = (3+t)^2 - 27$

g. $f(x) = x^4 + 2x^2 - 31x + 4$

h. $f(t) = \sqrt{t} - 5t^2$

i. $g(x) = \frac{(x-4)^2}{x}$

17. The rule at right is a combination of the Power Rule and what rule in Model 3?

18. (Check your work) Check that this combined rule is consistent with your answer to Question 13.

Summary Box DT1.2: Combined Rule

$$\frac{d}{dx}\left(cx^n\right) = cnx^{n-1}$$

where c and n are real numbers

19. Find the derivative of each function.

a. $f(x) = x^3$

b. $f(x) = 3x^2$

c. $f(x) = 6x$

20. Consider the functions in the previous question:

a. Construct an explanation for why $f(x) = 6x$ (in part c, above) is called the **second derivative** of the function $f(x) = x^3$ (in part a, above).

b. What is the **third derivative** of $f(x) = x^3$? Explain your reasoning.

c. (Check your work) Are your answers above consistent with the fact that the **fourth derivative** of $f(x) = x^3$ is zero?

Summary Box DT1.3: Notation for Second, Third, and Higher Derivatives

For a function $f(x)$...

- (review) the **first derivative** is represented as $f'(x)$
- the **second derivative** is represented as $f''(x)$
- the **third derivative** is represented as $f'''(x)$
- the ***n*^th^ derivative** is represented as $f^{(n)}(x)$ (where $n \geq 4$)
  For example, $f^{(4)}(x)$ is used, not $f''''(x)$.

21. Go back to Question 19 and, as shown in Summary Box DT1.3, add one or more marks (′) to the f in parts b and c to indicate each function's relationship with the function $f(x) = x^3$.

22. Find $f^{(4)}(x)$ for each of the following functions. Use the notation in Summary Box DT1.3 to indicate the first through third derivatives that you found along the way.

    a. $f(x) = x^5$

    b. $f(x) = x^8 - 5x^3 + 3x - 1$

Notes

Notes

DT2: Product and Quotient Rules

Construct Your Understanding Questions

1. Use the rules you learned in the previous activity (**Power, Constant Multiple, Sum** and **Difference Rules**) to find the derivative of each function in Model 1, below. Note: it may help to rewrite the function in a form that allows you to use these rules. Show any work.

Model 1: Finding Derivatives

	$f(x)$	*rewrite* $f(x)$ *if necessary*	$f'(x)$
i.	$f(x) = x - 1$		$f'(x) =$
ii.	$f(x) = (3x+1)(5x-2)$		$f'(x) =$
iii.	$f(x) = (4x^2)(5x^3)$		$f'(x) =$
iv.	$f(x) = \dfrac{7x^2 - 1}{x}$		$f'(x) =$

Construct Your Understanding Questions

2. (Check your work) **Functions ii-iv** in the table above must be rewritten to use the rules we have learned thus far to find the derivative. For example, **Function iv** can be written as $f(x) = 7x - x^{-1}$. Go back and check that you rewrote these three functions before you found their derivatives.

3. Steve's derivatives (along with an intermediate step) are shown below. The instructor looks at these answers and says "Please explain the method you used to generate these derivatives."

ii.	$f(x) = (3x+1)(5x-2)$	$f'(x) = (3)(5) = 15$
iii.	$f(x) = (4x^2)(5x^3)$	$f'(x) = (8x)(15x^2) = 120x^3$
iv.	$f(x) = \dfrac{7x^2-1}{x}$	$f'(x) = \dfrac{14x}{1} = 14x$

   a. Do Steve's derivatives match your derivatives from Question 1?

   b. Steve's intermediate step gives insight into his method. From this, reconstruct the rule he seems to be trying to apply here. Does this rule appear to work? Explain.

c. (Check your work) Cross out Steve's derivatives for **Functions ii-iv** at the bottom of the previous page so you remember these are wrong and his method is invalid.

4. It turns out that there *is* a rule (called the Product Rule) which *will* give the correct derivative for functions such as **Functions ii-iv** without having to multiply them out. Use your answers to Question 1 to determine which of the following is the *Real* Product Rule?

a. $\frac{d}{dx}[f \cdot g] = f' \cdot g'$

b. $\frac{d}{dx}[f \cdot g] = 2x \cdot f' \cdot g' - 1$

c. $\frac{d}{dx}[f \cdot g] = x\left(f' \cdot g' - 20x^3\right)$

d. $\frac{d}{dx}[f \cdot g] = g \cdot f' + f \cdot g'$

5. Apply the rule you chose in the previous question to find the derivative of **Function ii**, $f(x) = (3x+1)(5x-2)$. Show your work.

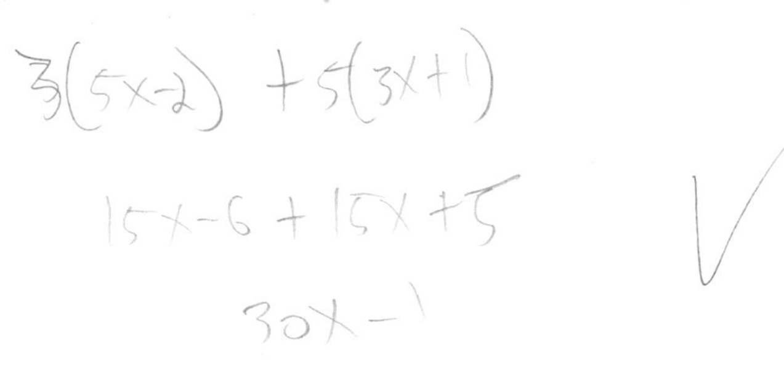

6. (Check your work) Does your answer in the previous question match your entry for the derivative of **Function ii** in Model 1? If the result does not match, you may have chosen the wrong rule in Question 4.

7. (Check your work) One of the choices in Question 4 represents Steve's "Invalid Product Rule." Which one? Go back and write: "Warning! Common Error! Do Not Use." next to this invalid rule

8. (Check your work) Are your answers on the previous page consistent with Summary Box DT2.1?

Summary Box DT2.1: Product Rule and Quotient Rule

If f and g are differentiable then

$$\frac{d}{dx}[f(x)\cdot g(x)] = g(x)\cdot\frac{d}{dx}[f(x)] + f(x)\cdot\frac{d}{dx}[g(x)]$$

and

$$\frac{d}{dx}\left[\frac{f(x)}{g(x)}\right] = \frac{g(x)\cdot\frac{d}{dx}[f(x)] - f(x)\cdot\frac{d}{dx}[g(x)]}{[g(x)]^2}\;; \qquad g(x)\neq 0$$

9. a. Apply the Quotient Rule to **Function iv** written in its original form: $f(x)=\dfrac{7x^2-1}{x}$. Show your work.

b. Apply the product rule to **Function iv** in Model 1 written as $f(x)=(x^{-1})\cdot(7x^2-1)$. Show your work.

c. (Check your work) Check that your answers to parts a and b match your answer to Question 1 for $\dfrac{d}{dx}[7x - x^{-1}]$.

10. In Column 2, write the derivative of the function f (show your work).
In Column 3, list the rule(s) you used to find this derivative.

	$f(x)$	$f'(x)$	Rules
v.	$f(x)=\frac{5}{x}$	$-\frac{5}{x^2}$	
vi.	$f(x)=\frac{2x^2-x}{x}$	2	
vii.	$f(x)=\frac{x+2}{x-5}$		
viii.	$f(x)=\frac{x^2-6x+9}{x-3}$		
ix.	$f(x)=(6x^2+3x+5)(x^3-2x^2-4)$		
x.	$f(x)=\frac{6\sqrt{x}}{3x+4}$		
xi.	$f(x)=\frac{x^2+x+1}{x^2-2}$		

11. The derivatives of **Functions v** and **vi** in the table above *can* be found using the Quotient Rule, but it is not the best choice.

a. Rewrite **Functions v** and **vi** as power functions (or a difference of power functions).

b. (Check your work) Find the derivatives of your rewritten functions. Do they match your table entries for these functions?

Use the graphs to answer the next series of questions. Note that h has a sharp corner at $x = 2$.

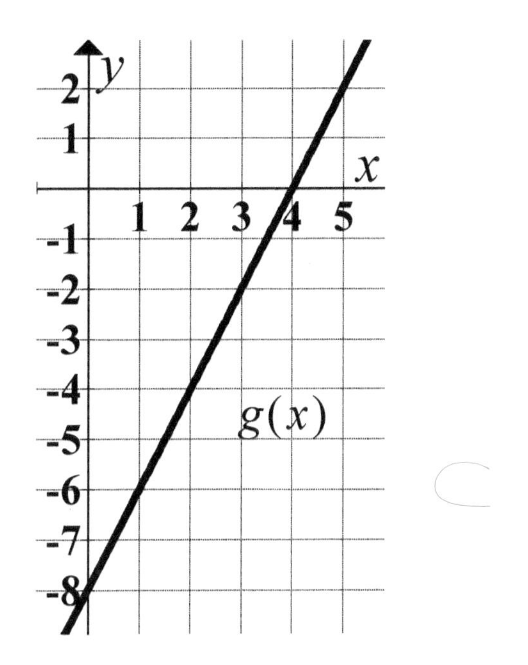

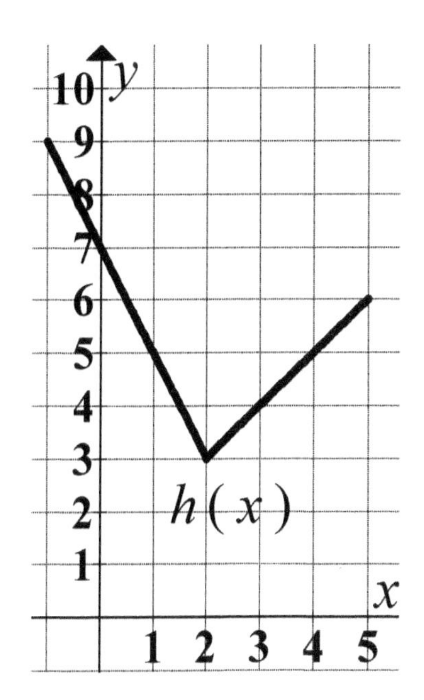

12. If $f(x) = g(x) \cdot h(x)$, find...

a. $f'(0)$ b. $f'(2)$ c. $f'(4)$

13. If $f(x) = \dfrac{g(x)}{h(x)}$, find...

a. $f'(0)$ b. $f'(2)$ c. $f'(4)$

14. If $f(x) = \dfrac{h(x)}{g(x)}$, find...

a. $f'(0)$ b. $f'(2)$ c. $f'(4)$

Notes

DT3: Derivatives of Exponential Functions

Model 1: Graph of $y = e^x$

The symbol e is used to represent the number whose nonrepeating decimal expansion begins: 2.71828182845904523536028747135266249775724709369995957496 6....

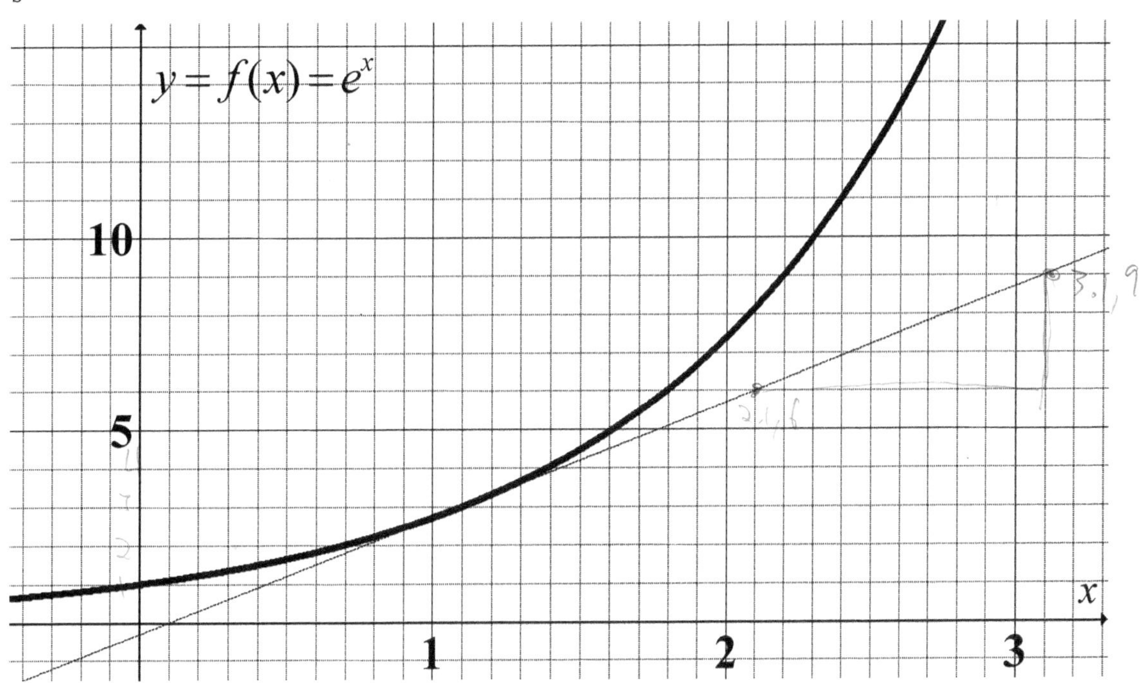

Construct Your Understanding Questions (to do in class)

1. According to Model 1, what do the numbers e and π have in common?
2. Use the graph of $f(x) = e^x$ in Model 1 to estimate the value of f at $x = 1.1$.

3. In Model 1, estimate the slope of the tangent line to the graph of $y = f(x)$ at $x = 1.1$ (shown).

4. Which question above is asking you to find $f(1.1)$, and which is asking you to find $f'(1.1)$? Explain your reasoning.

Model 2: $f(x)=c^x$ for Various Values of c

Note that decimals have been rounded to the thousandths position.

c	x	$f(x)=c^x$	$f'(x)$	$\frac{f'(x)}{f(x)}$
2	1	2	1.386	.693
2	2	4	2.773	.693
2	3	8	5.545	.693
2	4	16	11.090	.693
e	1	2.718	2.718	1
e	1.1	3.004	3.004	1
e	2	7.389	7.389	1
e	3	20.086	20.086	1
e	4	54.598	54.598	1
3	1	3	3.296	1.09
3	2	9	9.888	1.09
3	3	27	29.663	1.09
3	4	81	88.988	1.09
4	1	4	5.545	1.386
4	2	16	22.181	1.386
4	3	64	88.723	1.36
4	4	256	354.891	1.386

Construct Your Understanding Questions (to do in class)

5. Why does it make sense to list $c=e$ on the table after $c=2$ and before $c=3$?

6. (Check your work) Are your answers to Questions 2 and 3 consistent with the corresponding table entries?

yes

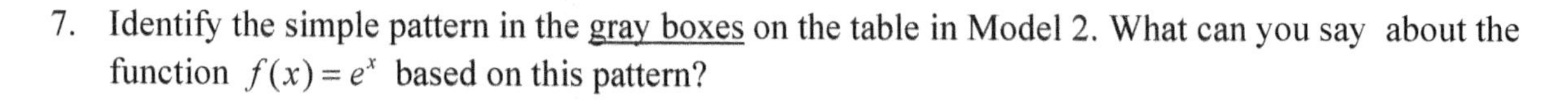

7. Identify the simple pattern in the gray boxes on the table in Model 2. What can you say about the function $f(x)=e^x$ based on this pattern?

8. (Check your work) For the generalized function $f(x)=c^x$ ...

   a. According to the table in Model 2, it appears that $f(x)=f'(x)$ when $c=$ e

   b. Does $f(x)=f'(x)$ for any of the other values of c listed on the table?

   no

9. Fill in the last column in Model 2 (use a calculator if you wish). Round your entries to the thousandths position.

   a. (Check your work) Did you find that when $c=2$, $\frac{f'(x)}{f(x)}=\ln 2$? yes

   b. Do other entries in the table (e.g., when $c=3$) follow this same pattern? If so, describe this pattern.

   yes

   $\frac{f'(x)}{f(x)}=\ln c$

10. Based on your answer to the previous question, devise an expression for each derivative.

    a. for $f(x)=2^x$, what is $f'(x)=$ $\frac{f'(x)}{f(x)}=\ln c$ $f(x)\cdot\ln c=f'(x)$ $f(x)\ln 3=f'(x)$

    b. for $f(x)=e^x$, what is $f'(x)=$ $f(x)\ln e$

    c. for $f(x)=3^x$, what is $f'(x)=$ $f(x)\ln 3$

    d. for $f(x)=4^x$, what is $f'(x)=$ $f(x)\ln 4$

11. (Check your work) Is your answer to part b of the previous question consistent with the fact that $\ln e=1$? Explain.

    yes

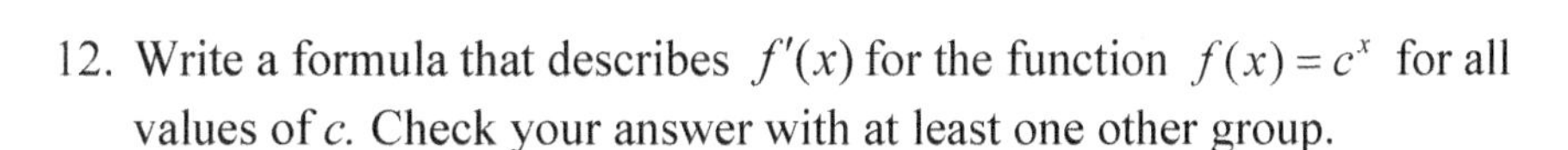

12. Write a formula that describes $f'(x)$ for the function $f(x)=c^x$ for all values of c. Check your answer with at least one other group.

13. Describe in words what you think is special about the number e. That is, why do you think this number is given a special symbol e?

    its derivative is itself

Notes

DT4 Pre-Activity: Review of Trigonometry

Complete this activity in preparation for CalcActivity DT4: Derivatives of Trigonometric Functions

Model 1: Circle of Radius = 1 (Unit Circle)

Consider the circle of radius 1 shown at right.

We will use the *Greek* letter θ to represent the angle between the positive part of the x axis and a ray, as shown by the curved arrow labeled θ.

θ can be described in degrees or by the part of the circumference (the arc length) defined by θ (shown in bold on the circle at right).

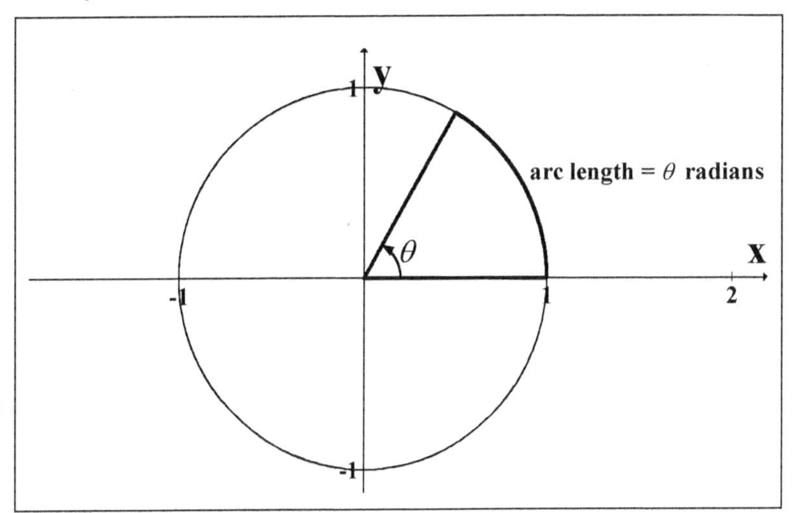

Construct Your Understanding Questions (to do in class)

1. (Review) In terms of π what is the circumference of the circle shown in Model 1? If you are uncertain of your answer, check your answer with another group.
2. (Review) Explain why, if $\theta = 360^\circ$, then the arc length associated with θ would be 2π.

3. (Check your work) When we use arc length to define an angle, we say the angle is measured in **radians** (rad). Is your answer above consistent with $\theta = \pi$ rad when $\theta = 180°$? yes
4. (Review) Fill in the missing values of θ <u>in radians</u> on the table below.

θ in degrees	0°	30°	45°	60°	90°	120°	135°	150°
θ in radians	0	$\frac{\pi}{6}$	$\frac{\pi}{4}$	$\frac{\pi}{3}$	$\frac{\pi}{2}$	$\frac{2\pi}{3}$	$\frac{3\pi}{4}$	$\frac{5\pi}{6}$

θ in degrees	180°	270°	360°	540°	720°	900°	1080°	1260°
θ in radians	π	$\frac{3\pi}{2}$	2π	3π	4π	5π	6π	7π

5. θ shown in Model 1 is equal to one of the values you just entered into the table, above. Without measuring, determine which is the most likely value of θ on the graph in Model 1?

$\frac{\pi}{3}$

Model 2: Sine, Cosine, and Tangent Functions

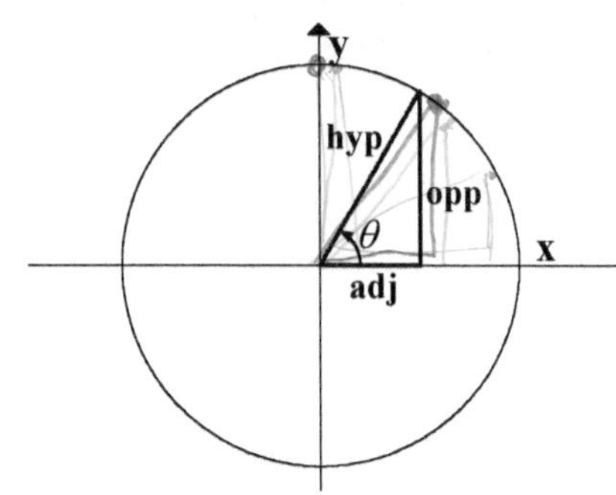

The functions sine, cosine, and tangent can be thought of as ratios of the sides of a right triangle inscribed in the upper right quadrant of a circle with the angle θ as marked at left.

$$\sin\theta = \frac{\text{opposite}}{\text{hypotenuse}} \qquad \cos\theta = \frac{\text{adjacent}}{\text{hypotenuse}}$$

$$\tan\theta = \frac{\text{opposite}}{\text{adjacent}}$$

Construct Your Understanding Questions (to do in class)

6. Use the triangles below (angles shown in radians), to determine the values of sin, cos, and tan in the table. (Hint: If you are stuck on the shaded column, read the next question.)

$\theta \rightarrow$	$\frac{\pi}{6}$	$\frac{\pi}{4}$	$\frac{\pi}{3}$	$\frac{\pi}{2}$	0	$\frac{2\pi}{3}$
$\sin\theta$						
$\cos\theta$						
$\tan\theta$						

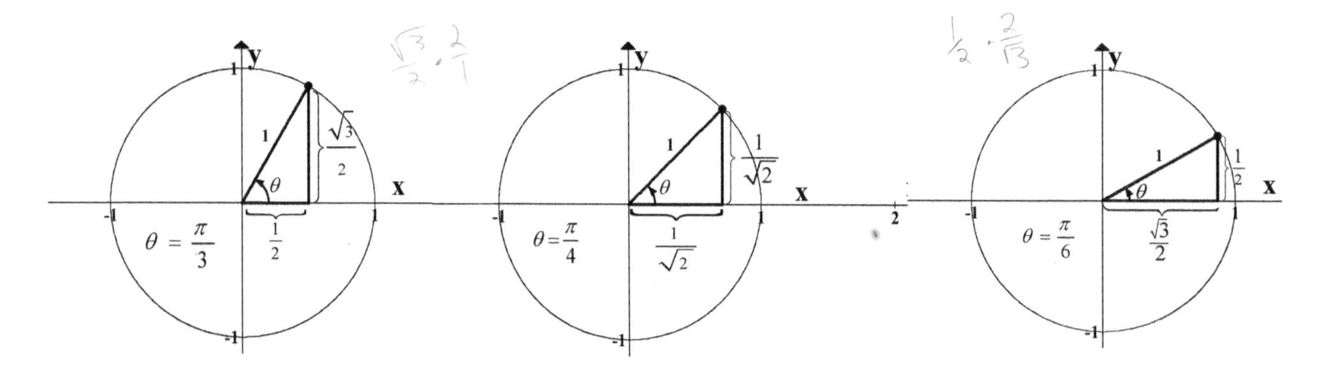

7. A general expression for sine, useful for angles greater than $\frac{\pi}{2}$, is $\sin\theta = \frac{y}{r}$. For $\theta = \frac{2\pi}{3}$: $y = \frac{\sqrt{3}}{2}$, $r = 1$. Find $\sin\frac{2\pi}{3}$.

8. Speculate: Where did these values of y and r come from? Hint: Write in the coordinates of the point marked on each circle in Question 6. (The circle at right is done for you.)

9. (Check your work) Does your answer explain why y = opp, and r = hyp = 1 for all the circles shown in Question 6?

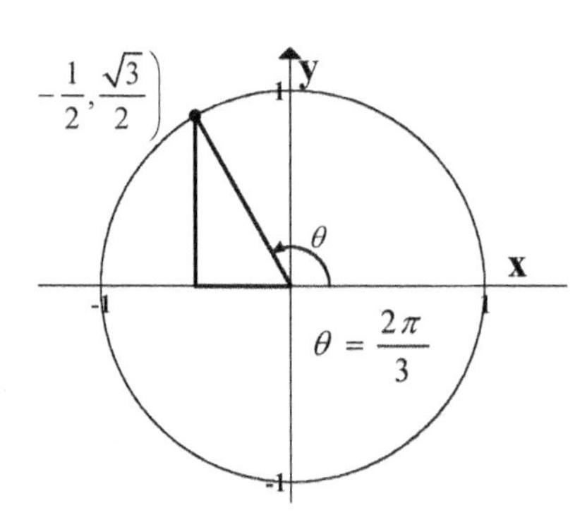

10. Use your conclusions in the previous questions to generate expressions for $\cos\theta$ and $\tan\theta$ that are comparable to the expression for $\sin\theta$ in the previous question ($\sin\theta = \dfrac{y}{r}$).

11. Use your expressions in the previous question to evaluate $\cos\dfrac{2\pi}{3}$ and $\tan\dfrac{2\pi}{3}$, and check your answers with at least one other group.

12. (Check your work) Are your answers to the previous questions consistent with the expressions shown at the top of Summary Box DT4.1? If not, go back and check your conclusions. Write down any questions and discuss them with another group or the instructor.

Summary Box DT4.1: Trigonometric Relations

For a circle centered at zero of radius r, where x and y are the coordinates of a point on the circle at to the intersection of a line segment starting at the origin and making an angle of θ (in radians) with the x-axis:

$$\sin\theta = \frac{y}{r} \qquad \cos\theta = \frac{x}{r} \qquad \tan\theta = \frac{y}{x}$$

Some other useful trigonometric relations (θ and ϕ are angles in radians):

$$\sin^2\theta + \cos^2\theta = 1 \qquad \tan\theta = \frac{\sin\theta}{\cos\theta} \qquad \sec\theta = \frac{1}{\cos\theta} \qquad \csc\theta = \frac{1}{\sin\theta}$$

Notes

DT4: Derivatives of Trigonometric Functions

Model 2: Graph of sin x

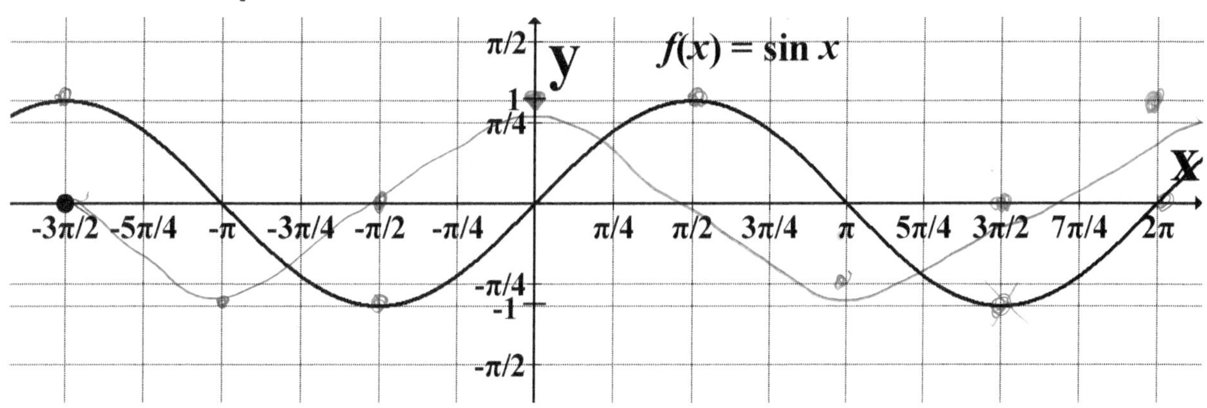

Unless noted, assume for all trigonometric functions the argument (e.g., x, θ, ϕ, etc.) is in radians.

Construct Your Understanding Questions (to do in class)

13. Confirm that $f(x)$ in Model 2 matches your entries in the table in Question 6 for $\theta = 0$, $\frac{\pi}{4}$, and $\frac{\pi}{2}$.

14. Begin plotting $f'(x)$ for f in Model 2 by plotting each point where $\frac{d}{dx}[\sin x] = 0$. One is done for you.

15. a. Use a graphical method to calculate $\left.\frac{d}{dx}[\sin x]\right|_{x=0}$ for the graph in Model 2. That is, find the slope of the tangent line to sin x at $x = 0$. Plot the point $(0, f'(0))$ on the graph.

    b. At what other value of x is $f'(x) = f'(0)$? That is, where is the slope of the tangent to f the same as at $x = 0$. Add this point to your graph of $f'(x)$ in Model 2.

    c. Estimate the slope of the tangent line to $\sin x$ at $x = \pi$ and $x = -\pi$. Plot these points $(\pi, f'(\pi))$ and $(-\pi, f'(-\pi))$ on the graph in Model 2.

    d. Do your best to sketch a smooth curve of the function $f'(x)$ on the graph in Model 2.

    e. The graph you just sketched looks like another trigonometric function. Can you guess which one it is? (Check your work) Look back at the table in Question 6 to confirm your guess.

Model 3: Graph of sin x and cos x

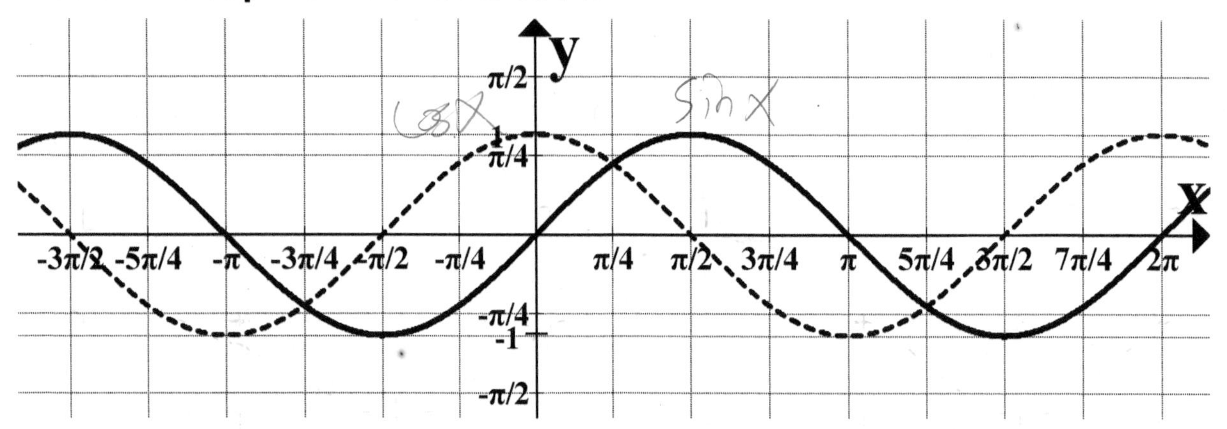

Construct Your Understanding Questions (to do in class)

16. There are two trigonometric functions shown in Model 3. One is dotted and one solid.

    a. Label one "sin x" and the other "cos x."

    b. When $f(x) = \cos x$, $f'\left(\frac{\pi}{2}\right) =$

    c. When $f(x) = \sin x$, $f'\left(\frac{\pi}{2}\right) =$

    d. $\sin\left(\frac{\pi}{2}\right) = 1$, and $\cos\left(\frac{\pi}{2}\right) = 0$ Based on this, which one of the following statements is true? $\frac{d}{dx}(\sin x) = \cos x$ or $\frac{d}{dx}(\cos x) = \sin x$ [Circle one]

17. f plotted below is $\cos x$. (Check) Is this consistent with your answer to Question 16a?

    a. On these same axes (below), plot $f'(x)$.

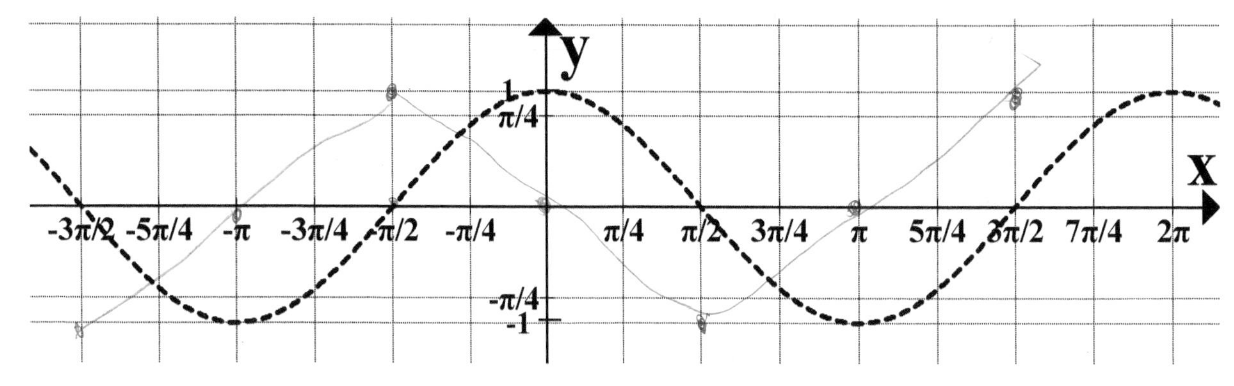

    b. Identify the function you plotted. That is, what is $\frac{d}{dx}(\cos x) =$

18. Which one of the following is a graph of $\frac{d}{dx}(\cos x)$?

(Check your work) Is this consistent with your sketch of $\frac{d}{dx}(\cos x)$ on the previous page?

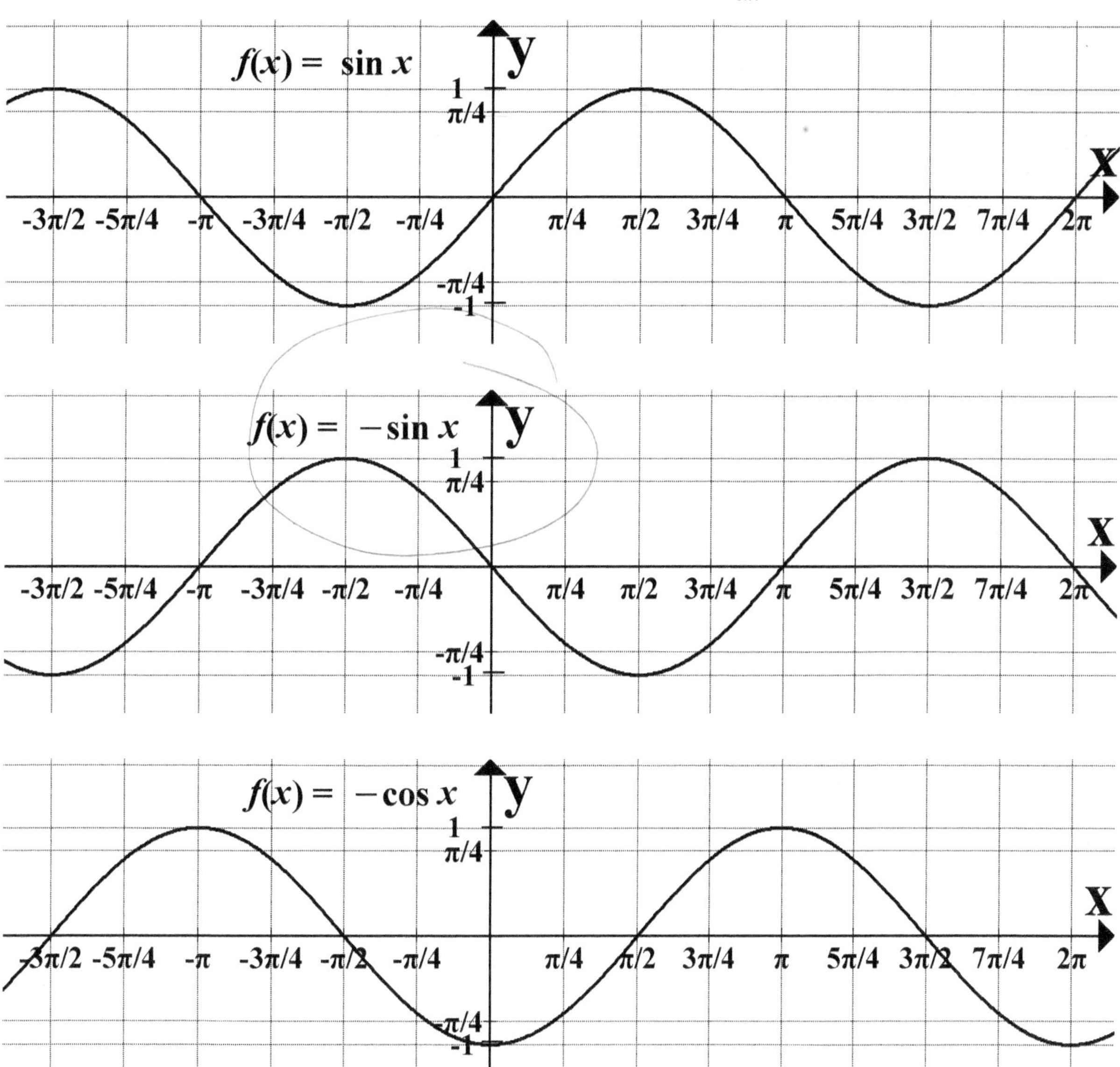

19. (Check your work) Are your answers above consistent with Summary Box DT4.2? If not, review and revise your answers on the previous three pages.

> **Summary Box DT4.2: Derivatives of Sine and Cosine**
>
> $\frac{d}{dx}(\sin x) = \cos x$ $\qquad\qquad$ $\frac{d}{dx}(\cos x) = -\sin x$

20. Use the quotient rule to derive an expression for $\frac{d}{dx}(\tan x)$. Hint: to simplify your result, use one of the trigonometric relations found in Summary Box DT4.1. When you are done, check your work with another source.

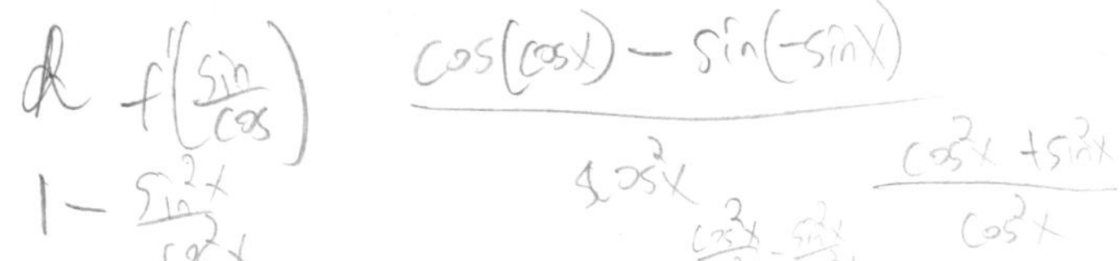

21. If $f(x) = \sin x$ then $f'(x) = \cos x$, but what are the second, third, fourth, etc. derivatives?

a. $f''(x) =$

b. $f'''(x) =$

c. $f^{(4)}(x) =$

d. $f^{(5)}(x) =$

e. $f^{(6)}(x) =$

22. Complete the sentence: If $f(x) = \sin x$ then $f^{(n)}(x) = \sin x$ when n is a multiple of _____. (You may want to extend the series in the previous question to higher values of n to be sure of your answer.)

Notes

Notes

DT5 Pre-Activity: Review of Compositions

This activity must be completed BEFORE attempting CalcActivity DT5: The Chain Rule

Model 1: Long Distance Road Race

$P(t)$ gives the position of a race car as a function of time during a 1000 km road race.

Graph I: Race Car Position

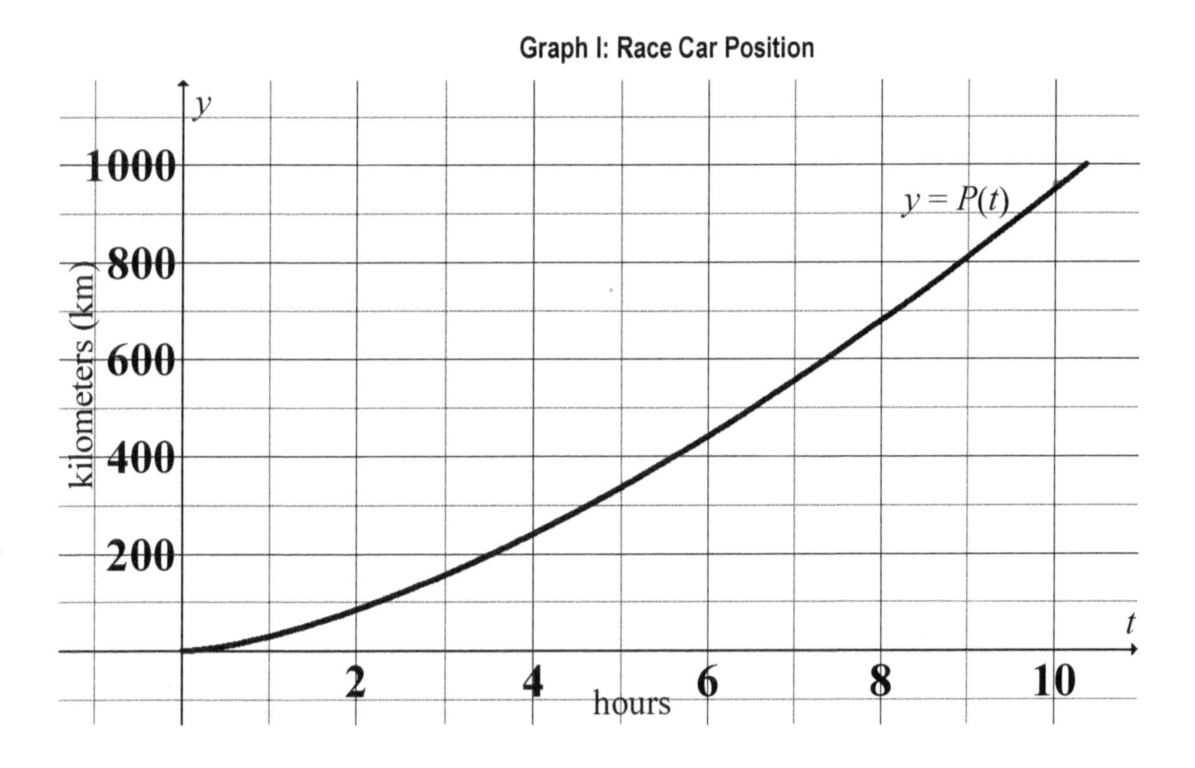

Construct Your Understanding Questions (to do in class)

1. At each point on the curve, what information about the car is conveyed by the slope of the line tangent to the curve in Graph I?

2. The race course has one section of curvy mountainous roads where drivers have to go relatively slowly. Is this mountainous section at the beginning or the end of the race? Cite evidence from Model 1 to support your answer.

3. Use the graph to estimate the instantaneous velocity of the car at $t = 10$ hours. Show your work and **include units in your answer**.

Model 2: Tire Wear

The car in Model 1 begins the race with new tires which have 10,000 μm (10mm) of tread. The function $W(u)$ in Graph II describes tire wear (change in tread thickness) during the race.

Graph II: Tire Wear as a Function of Distance

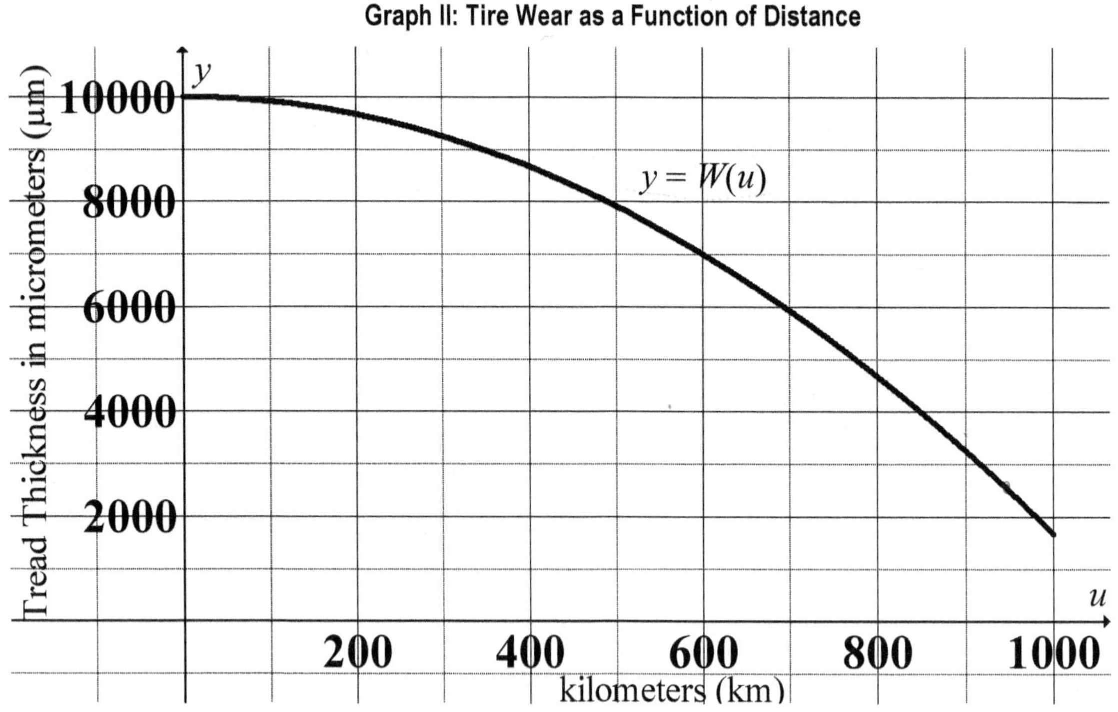

Construct Your Understanding Questions (to do in class)

4. The tires in Models 1 and 2 tend to fail if they have less than 1 mm of tread. Is the car expected to finish the race without having to change tires? Explain your reasoning.

5. Are the tires on the racecar losing rubber (wearing out) faster near the start of the race or near the end of the race? Cite evidence from Model 2 to support your answer.

6. Use the graph in Model 2 to estimate the slope of the tangent line at $u = 950$. Show your work **including units**. Describe what this value tells you about the racecar at $u = 950$ km.

7. Which notation best describes the quantity you calculated in in the previous question?

u, $W(u)$, $W'(u)$, t, $P(t)$, $P'(t)$, $W(950)$, $W'(950)$, $P(10)$, $P'(10)$, $W(950)-W(0)$, $\dfrac{W(u)}{u}$, $\dfrac{W(950)-W(0)}{950}$

8. Complete the table by entering appropriate notation from Question 7 for each row.

Description	Value	Notation from Above
Total number of miles traveled at t = 10 hours	950 km	p(10)
Car velocity at t = 10 hours	142 km/hour	p'(10
Total change in tread thickness during the first 950 km	-7500 μm	w(950)−w(0)
Average rate of change of tread thickness over the first 950 km	-8 μm/km	$\frac{w(950)-w(0)}{950}$
Instantaneous rate of change in tread thickness at 950 km	-16 μm/km	w'(950)

9. (Check your work) The table above contains the answer to Question 6. Make sure this is consistent with your answers, and mark that row in the table "Answer to Question 6."

10. On the graph below, plot the tire thickness at t = 0, 2, 4, 6, 8, and 10 hours.

Graph III: Tire Wear as a Function of Time

Tread Thickness in micrometers (μm)

y

10000
8000
6000
4000
2000

y = w∘p(t)

2 4 6 8 10 t

hours

11. Write a description of how to find a y value for a given time (t) on Graph III. Use $t = 10$ as an example. That is, explain how you found the point for $t = 10$ hours on Graph III.

12. Write a general set of instructions for finding a point on Graph III. Include each of the following notations at least once in your instructions: u, $W(u)$, t, and $P(t)$.

13. Which of the following notations best describes a point on Graph III? Circle one.

    (t,u) $(t,W(t))$ $(t,P(t))$ $(t,P(W(t)))$ $(t,W(P(t)))$

    Explain your reasoning.

14. (Check your work) One of the following compositions describes the function you plotted on Graph III. Circle one, and write "$y =$ [your answer]" on Graph III.

    $(W \circ P)(t)$ or $(P \circ W)(t)$

15. (Check your work) The convention for writing the composition $f \circ g$ is represented by the cartoon below.

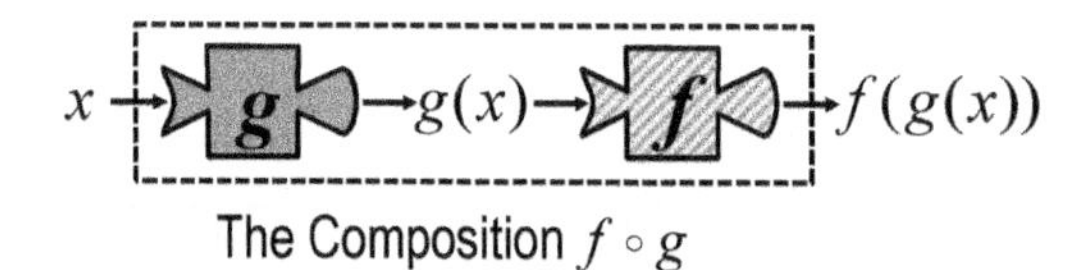

The Composition $f \circ g$

   a. Is this notation convention consistent with your answer to the previous question?

   b. Draw a cartoon modeled on the one above that represents the composition you circled in Question 14.

   c. (Check your work) A student says that the following cartoon is equivalent to the one you drew. Agree or disagree and explain your reasoning.

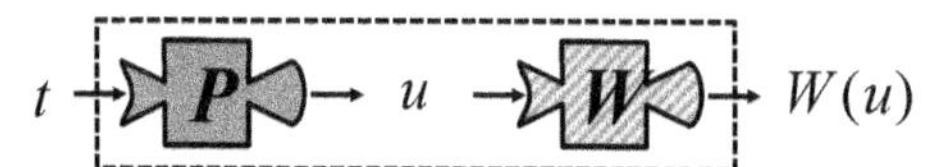

Differentiation Techniques 5: The Chain Rule

Model 3: Models 1 and 2 Revisited

Position Function from Graph I

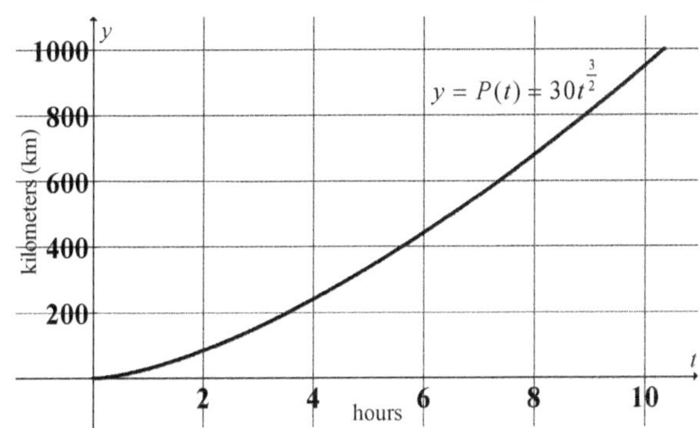

Race Car Position as a Function of Time

Tire Wear Function from Graph II

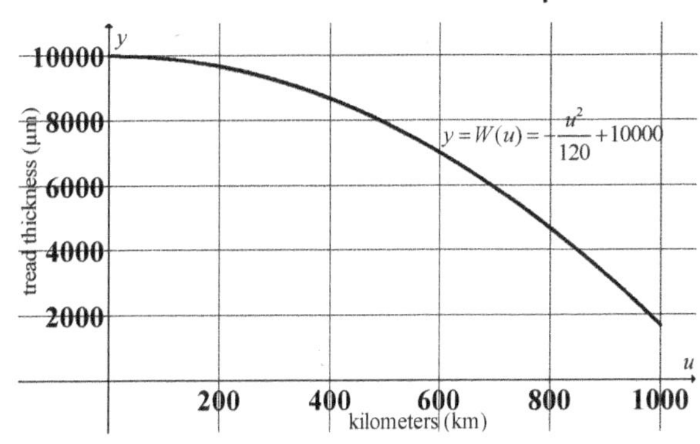

Tire Tread Thickness as a Function of Distance.

Construct Your Understanding Questions (to do in class)

16. Find values for … (Note: Graphs I and II are larger in Models 1 and 2.)

    a. $P(10) =$

    $30(10)^{\frac{3}{2}} = 948$

    b. $W(P(10)) =$

    $-\frac{u^2}{120} + 1000$

    2510

17. In Question 3 you found that $P'(10)$ was close to 142.

    a. Describe what $P'(10) = 142$ tells you about the racecar.

    Speed at 10 hours

    b. What are the units of $P'(t)$?

    Km/hr

18. In Question 6 you found that $W'(950)$ was close to -15.8.

    a. Describe what $W'(950) = -15.8$ tells you about the racecar.

    loses 16 μm per Km at 950

    b. What are the units of $W'(u)$?

    μm/km

19. Calculate the following product (**include units in your answer**):

$P'(10) \cdot W'(950) =$ −2243

a. Which best describes what this product tells you about the car?

i. tire wear per kilometer at $u = 950$ km

ii. tire wear per hour at $t = 10$ hrs

iii. tire wear at a speed of 142 km/hour

Circle one, and **explain your reasoning**.

b. All but one of the following is a generalized form of the product $P'(10) \cdot W'(950)$, above. Cross out the one that is not correct.

$P'(t) \cdot W'(P(t))$, ~~$P'(t) \cdot W'(t)$~~, $P'(t) \cdot W'(u)$

20. (Check your work) The product that you calculated in Question 19 (−2244 μm/hr) is equal to the slope of a tangent line shown on Graph III, below, right.

a. Identify which tangent line.

b. Which of the following is equal to −2244 μm/hr?

i. $(W \circ P)(10)$

ii. $(W \circ P)'(10)$

iii. ~~$(W \circ P)(950)$~~

iv. $(W \circ P)'(950)$

Explain your reasoning.

21. (Review) The derivative of a composition of two functions, $f(u)$ and $g(x)$ where $u = g(x)$, can be written: $(f \circ g)'(x)$.

a. What are the units of $(W \circ P)'(10)$, and what does this value tell you about the racecar?

u/h

b. Write an expression for the **derivative** of the function in Graph III, $(W \circ P)'(t)$ using some but not all of the symbols: P, W, P', W', t, or u. Hint: Make sure the units of the functions and variables you choose match the units of $(W \circ P)'$.

$(W \circ P)'(t) =$

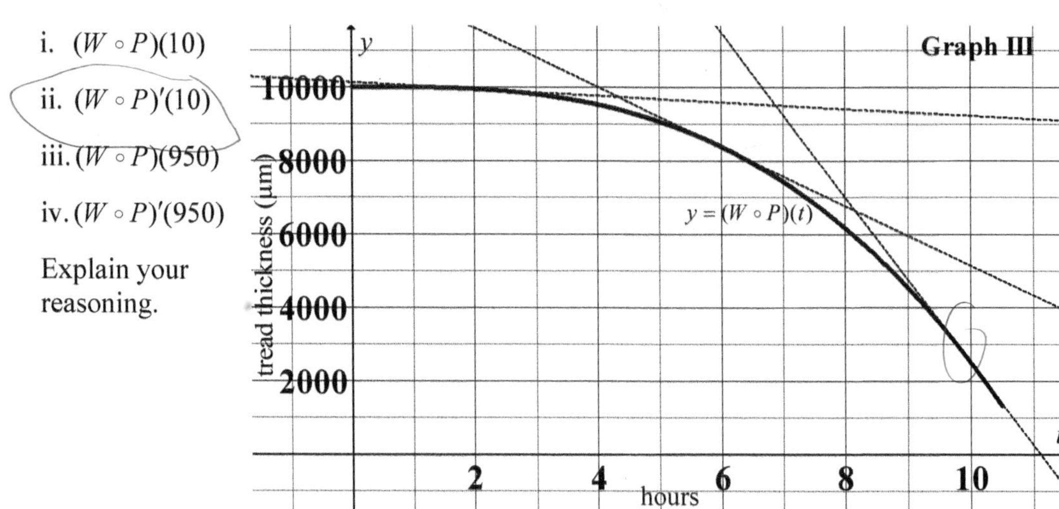

22. (Check your work) There is a parallel between Question 19 and Question 21. Check that your answers reflect this parallel, and discuss this parallel with your group. If you do not see a parallel, consult with another group or the instructor.

23. Recall the functions $P(t) = 30t^{\frac{3}{2}}$ and $W(u) = -\frac{u^2}{120} + 10000$ from Model 3.

a. Which one of the following is a simplified expression for $(W \circ P)(t)$?

$-\frac{30t^3}{120} + 10000$ $\quad -\frac{15t^3}{2} + 10000$ $\quad (-\frac{u^2}{4} + 300000)^{\frac{3}{2}}$ $\quad -\frac{t^3}{4} + 10000$

b. Find the derivative with respect to time (t) of the equation you chose in part a.

c. Find $P'(t)$.

d. Find $\frac{d}{du}W(u)$.

Be sure to simplify your answers.

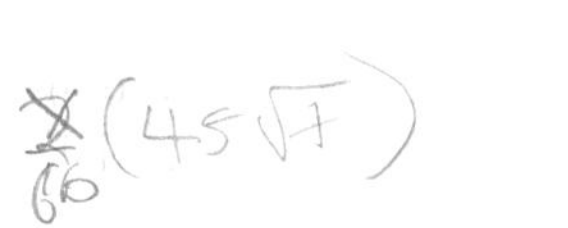

e. Which of the following is a simplified expression for $W'(P(t))$?

$-\frac{1}{2}t^{\frac{3}{2}}$ $\quad -\frac{15t^3}{2} + 10000$ $\quad -\frac{3\sqrt{t}}{4}$ $\quad 45\sqrt{t}$ $\quad -\frac{t}{60}$

f. (Check your work) Substitute $u = 30t^{\frac{3}{2}}$ into your answer in part d. If this does not match your expression for $W'(P(t))$ then revisit your answers to d and e.

24. In Question 20 you were told that $(W \circ P)'(10) = -2244$ μm/hr. This means that at $t = 10$ hours the car is losing 2244 μm (or 2.244 mm) of tread per hour. In Question 19 you found this same value by multiplying two numbers together.

a. In Question 19, what two numbers were multiplied together to find $(W \circ P)'(10)$?

b. What two expressions in Question 23, above, can be multiplied together to give your result in Question 23b: $(W \circ P)'(t)$?

c. (Check your work) Multiply the two expressions you chose in part b. If they do not equal $-\frac{45t^2}{2}$, go back and revisit your answers on the previous page.

d. (Check your work) Are your answers to Questions 19b, 21b, and 24b consistent with Summary Box DT5.1?

Summary Box DT5.1: The Chain Rule

If $F = f \circ g$, that is, $F(x) = f(g(x))$, then $F'(x) = f'(g(x)) \cdot g'(x)$

provided that g is differentiable at x, and f is differentiable at $g(x)$.

Extend Your Understanding Questions (to do in or out of class)

25. For the functions P and W in Models 1-3, which of the following notations describe the quantity $P'(10) \cdot W'(950)$ that you calculated in question 19? Circle all that apply and explain your reasoning for each.

a. $\left.\frac{dW}{du}\right|_{u=950} \times \left.\frac{dP}{dt}\right|_{t=10}$

b. $(P \circ W)'(950)$

c. $(W \circ P)'(10)$

26. Decompose each function $F(x)$ by rewriting it as a composition of two functions $f(u)$ and $g(x)$ such that $F(x) = f(g(x))$.

a. $F(x) = (3x^2 + 20)^{100}$

b. $F(x) = 5\cos(27\sqrt{x})$

c. $F(x) = e^{5x^3 - 150}$

27. Use the Chain Rule to find the derivative of each of the starting functions F in the previous question.

28. Find the derivative of each function.

a. $f(x) = 5(x^2 - 10)^{14}$

b. $f(x) = (x^2 - 10)^2$

c. $f(x) = x^3 \sin(x)$

d. $f(x) = -3\cos(5x^3 + 7x)$

e. $f(x) = 15x^4 e^x$

29. For which functions in the previous question did you use the chain rule to find the derivative? For which functions did you have to use the chain rule? Explain why.

30. Give an example, not appearing in this activity, of a function that requires the chain rule to find the derivative, and explain your reasoning.

Notes

DT6: Derivatives of Inverse Functions

Construct Your Understanding Questions (to do in class)

1. Complete Rows 1 and 2 on the table in Model 1 by writing in the two missing compositions in the white boxes, then simplifying each and writing the result in the corresponding gray box (with the heading **simplified**).
2. Enter one of the following for $(g \circ g^{-1})(x)$ and $(g^{-1} \circ g)(x)$ in Row 3. Choose from:

$$x^{\sqrt{2}} \qquad \sqrt{x^2} \qquad (\sqrt{2})^x \qquad 2^{\sqrt{x}} \qquad (\sqrt{x})^2 \qquad \sqrt{2^x}$$

Model 1: Inverse Functions

Recall that the inverse function of $g(x)$, if it exists, is written $g^{-1}(x)$. See table below.

	$g(x)$	$g^{-1}(x)$	$(g \circ g^{-1})(x)$	→ simplified	$(g^{-1} \circ g)(x)$	→ simplified
Row 1	$x+4$	$x-4$			$(x+4)-4$	x
Row 2	$\frac{1}{2}x$	$2x$	$\frac{1}{2}(2x)$			
Row 3	x^2 $(x \geq 0)$	$\sqrt{x}$ $(x \geq 0)$				
Row 4	$(2x+8)^3$	$\frac{\sqrt[3]{x}-8}{2}$	$\left[2\left(\frac{\sqrt[3]{x}-8}{2}\right)+8\right]^3$		$\frac{\sqrt[3]{(2x+8)^3}-8}{2}$	
Row 5	e^x	$\ln x$ $(x>0)$	$e^{\ln x}$ $(x>0)$		$\ln(e^x)$	

Construct Your Understanding Questions (to do in class)

3. Complete Rows 3-4 on the table in Model 1, including the gray boxes labeled **simplified**. (Use scratch paper if necessary to convince yourself of the pattern in the gray boxes.)

   a. What is the result when you compose a function and its inverse?

   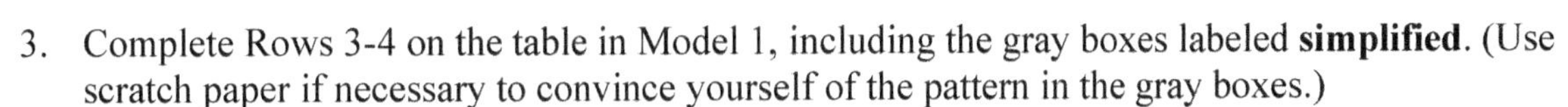

   b. (Check your work) Is your answer to part a consistent with the fact that the phrase "do and undo" is often applied to a function and its inverse? That is, whatever a function does to an input (e.g., x), the inverse function undoes it.

   c. Based on the pattern established by Rows 1-4, complete the table in Model 1 by filling in the gray boxes in Row 5.

4. One entry in Row 5 of Model 1 leads to the equation: $\frac{d}{dx}\left[e^{\ln x}\right]=1$

 a. Cite which entry and explain your reasoning.

 b. Use the **chain rule** to find the derivative of $e^{\ln x}$. The result will contain a term $\frac{d}{dx}[\ln x]$. Do not try to simplify this term. Just keep it as part of your equation. Show your work

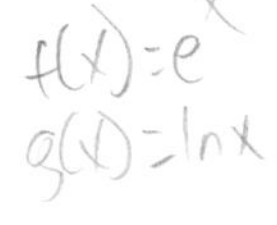

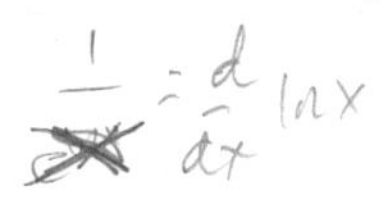

 c. Solve your equation in part b for $\frac{d}{dx}[\ln x]=$

5. (Check your work) If you have not already done so, replace $e^{\ln x}$ with x in your expression for $\frac{d}{dx}[\ln x]$. What is your new expression for: $\frac{d}{dx}[\ln x]=$ Check your result with another group.

6. The technique you just used to find the derivative of $\ln x$ can be used to find the derivative of any function that has an inverse, for example, for $f(x)=\sqrt{x}\ \ (x \geq 0)$

 a. What is the inverse of $\sqrt{x}$? That is, when $f(x)=\sqrt{x}$, what is $f^{-1}(x)=$

 b. (Check your work) Is $f^{-1}(f(x))=x$? If not, revise your answer to part a.

 c. For $f(x)=\sqrt{x}$ and $f^{-1}(x)=$ [your answer to part a of this question], which choice in Question 2 is equal to $f^{-1}(f(x))=$

 d. <u>Without using the power rule</u>, use the chain rule to find the derivative of your expression from part c. Be sure to preserve the term $\frac{d}{dx}\sqrt{x}$ in your equation.

e. Set your expression from part d (containing the term $\frac{d}{dx}\sqrt{x}$) equal to 1, and explain how we know this is expression is equal to 1.

f. (Check your work) Solve the equation in part e for $\frac{d}{dx}\sqrt{x} =$

g. Show that your answer to part f matches the result you get from using the power rule to find $\frac{d}{dx}\sqrt{x} =$

7. For Summary Box DT6.1 …

 a. Identify the line where the chain rule was used.

 b. Explain the logic behind Line 4.

 c. Use the result in Line 5 to find the derivative of $f(x) = x^{\frac{1}{3}}$. Show your work.

Summary Box DT6.1: Derivation of the General Formula for the Derivative of an Inverse Function

1. $f(f^{-1}(x)) = x$

2. $\frac{d}{dx}\left[f(f^{-1}(x))\right] = 1$

3. $\frac{d}{dx}\left[f(f^{-1}(x))\right] = f'(f^{-1}(x)) \cdot \frac{d}{dx}\left(f^{-1}(x)\right)$

4. $f'(f^{-1}(x)) \cdot \frac{d}{dx}\left(f^{-1}(x)\right) = 1$

5. $\boxed{\frac{d}{dx}\left(f^{-1}(x)\right) = \frac{1}{f'(f^{-1}(x))}}$

8. Fill in the blanks to complete Summary Box DT6.2. If your group is uncertain of any of your answers, be sure to check them with a resource such as another group, your instructor, or a textbook.

Summary Box DT6.2: Key Conclusions of this Activity (and some review)

e^x and __________ are inverses of each other

$\frac{d}{dx}[\ln x] =$ ____________________

$\frac{d}{dx}\left[e^x\right] =$ ____________________

$\frac{d}{dx}\left[c^x\right] =$ ____________________ where c is a constant

x^2 and __________ are inverses of each other $(x \geq 0)$

__________ and $x^{\frac{1}{n}}$ are inverses of each other (n = any integer. For n even, $x \geq 0$)

$\frac{d}{dx}[x^{\frac{1}{n}}] =$ ____________________

9. If c is a constant (not equal to e) then $\ln x$ and c^x are <u>not</u> inverse functions (that is, $\ln c^x = x$ only when $c = e$). However, $\ln c^x = x \cdot \ln c$ for all values of c.

   Use $\ln c^x = x \cdot \ln c$ and your formula for $\frac{d}{dx}[\ln x]$ in Summary Box DT6.2 to show that $\frac{d}{dx}\left[c^x\right] = c^x \cdot \ln c$. Show your work. (Try the problem without the hint on the next page.)

Hint: Take the derivative of both sides of $\ln c^x = \ln c \cdot x$, and preserve the term $\frac{d}{dx}\left[c^x\right]$.

Notes

DT7: Implicit Differentiation

Model 1: Solving for y

Most of the functions we have seen in this course are like those in Table 1 (and the first three functions in Table 2): They can be written **explicitly** in terms of x. That is, the function can be written as a single equation in the form $y = \ldots$ or, alternatively, $f(x) = \ldots$ in terms of x only.

Table 1
$y = 2x + 4$
$y = x^3 - 3x^2 + 13x - 7$
$f(x) = \dfrac{x+2}{x-2}$
$y = \sin x$
$f(x) = x^{1-\cos x}$

Table 2	Same function solved for y in terms of x only
$y - x^2 = x^3 + x$	
$18x + 5y = 12x^2 - y + 6$	
$xy - 7x = \sin x$	
$x^2 + xy + y^2 = 12$	
$x^2 y + y^2 + 7 = 0$	

Construct Your Understanding Questions (to do in class)

1. The first three functions in Table 2 can be written as explicit functions of x. Solve these for y, and write the new equations **explicitly** in the form $y = \ldots$ on the right side of Table 2. (Note: The last *two* equations in Table 2 cannot be uniquely solved for y in terms of x, so these boxes are darkened.)

2. Take the derivative of the equation you wrote in Row 1 of Table 2.

   $f'(x) =$

3. Recall that an alternate symbol for $f'(x)$ is y'.

   a. Which one of the following is also an alternate symbol for $f'(x)$?

   $\dfrac{dy}{dx}$ or $\dfrac{dx}{dy}$ Circle one, and explain your reasoning.

b. (Check your work) Is your answer to part a of this question consistent with the following? $\frac{d}{dx}[y] = y'$

4. (Check your work) In Row 1 of Table 2 you should have written the equation $y = x^3 + x^2 + x$ (or equivalent). **One way to think about taking the derivative of this equation is to take the derivative of each term on both sides of the equation.** This gives...

$$\frac{d}{dx}[y] = \frac{d}{dx}[x^3] + \frac{d}{dx}[x^2] + \frac{d}{dx}[x]$$

a. Find the derivative of $y = x^3 + x^2 + x$ by replacing each term in the derivative expression above with either y', or an expression in terms of x.

b. (Check your work) Check that your answer to part a) is the same as your answer to Question 2.

5. Alternatively, we can take the derivative of this function as it is written on the left side of Table 2: $y - x^2 = x^3 + x$. That is, without solving for y first, we can write...

$$\frac{d}{dx}[y] - \frac{d}{dx}[x^2] = \frac{d}{dx}[x^3] + \frac{d}{dx}[x]$$

a. Replace each term in the derivative expression above with either y', or an expression in terms of x.

b. (Check your work) Solve for y', then check that your equation is equivalent to your answers to Questions 2 and 4.

6. Find the derivative of $18x + 5y = 12x^2 - y + 6$ (found in Row 2 of Table 2) without first solving for y. That is, take the derivative of each term, then solving for y'. Show your work. Hint: Recall that $\frac{d}{dx}[5y] = 5 \cdot \frac{d}{dx}[y]$

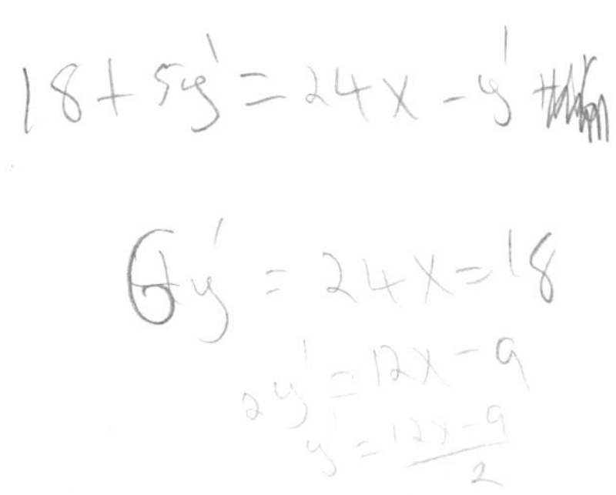

7. (Check your work) Take the derivative of the equation $y = 2x^2 - 3x + 1$ (which you should have written in Row 2 of Table 2), and make sure this matches your answer to the previous question.

$y = 4x - 3$

8. Look back at your entry in Row 3 of Table 2.

   a. (Check your work) Did you write $y = \dfrac{\sin x}{x} + 7$ or an equivalent expression?

   b. The following shows one way to find the derivative of this expression. What derivative rule is shown being used below to find this derivative?

$$y = x^{-1}\sin x + 7$$
$$y' = -x^{-2}\sin x + x^{-1}\cos x$$
$$y' = -\frac{\sin x}{x^2} + \frac{\cos x}{x}$$

9. Alternatively, we can take the derivative of this function as it is written on the left side of Table 2: $xy - 7x = \sin x$. That is, without solving for y first, we can write…

$$\frac{d}{dx}[xy] - \frac{d}{dx}[7x] = \frac{d}{dx}[\sin x]$$

   a. Two of the three terms in the derivative expression above depend on x only. Find the derivatives of these two terms.

   b. What derivative rule should be used to find the derivative of the remaining term? Explain your reasoning.

   c. (Check your work) Use the product rule to write an expression for $\dfrac{d}{dx}[xy]$. Represent $\dfrac{d}{dx}[y]$ using the symbol y', as you have done in previous questions.

10. (Check your work) Check that your answer to part c of the previous question is consistent with the following expression of the product rule: $\frac{d}{dx}[f \cdot g] = f \cdot \frac{d}{dx}[g] + g \cdot \frac{d}{dx}[f]$

11. Find the derivative of $xy - 7x = \sin x$ without first solving for y. That is, by taking the derivative of each term with respect to x then solving for y'. Show your work.

12. (Check your work) Becky and Milo get different answers to the previous question. Neither answer looks like the expression for y' in Question 8b, yet when they check with another group, the consensus is that all three answers are correct!

Becky's Answer	Milo's Answer	Result from Question 8b
$y' = \frac{7 - y + \cos x}{x}$	$y' = \frac{7 - \left(\frac{\sin x}{x} + 7\right) + \cos x}{x}$	$y' = -\frac{\sin x}{x^2} + \frac{\cos x}{x}$

a. Does your answer to the previous question match one of these answers? If not, go back and check your work.

b. What is the difference between Becky's answer and Milo's answer?

c. Explain why Milo was able to legitimately replace the y in Becky's answer with the expression shown in brackets in his answer.

d. Simplify Milo's answer so that it looks like the answer shown in Question 8b.

13. Now we will find the derivative of the fourth function in Table 2: $x^2 + xy + y^2 = 12$. Recall that this function cannot be uniquely solved for y in terms of x.

$$\frac{d}{dx}[x^2] + \frac{d}{dx}[xy] + \frac{d}{dx}[y^2] = \frac{d}{dx}[12]$$

a. Using the technique that you have learned in this activity, replace each term in the expression above with its derivative. (Circle any terms that are giving you trouble.)

b. (Check your work) Many students are not sure how to replace the term $\frac{d}{dx}[y^2]$. The next two questions are designed to help with this term.

14. Which of the following is equal to $\frac{d}{dx}[y^2]$? (more than one answer may be correct)

i. $y \cdot y' + y \cdot y'$ ii. $y \cdot y'$ iii. $2y \cdot y'$

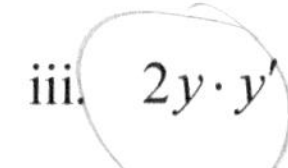

Explain your reasoning.

15. (Check your work) Check your choice in the previous question using the function, $y = 3x$ by giving each of the following answers in terms of x.

a. If $y = 3x$, then $y^2 =$

b. If $y^2 = 9x^2$, then $\frac{d}{dx}[y^2] =$

c. If $y = 3x$, then $y' =$

d. Substitute $y = 3x$ and $y' = 3$ into the equation you chose in Question 14, and check that this gives the same result as in part b of this question.

16. Continue your work from Question 13: Find the derivative of each term in the equation $x^2 + xy + y^2 = 12$, and then solve for y'. (Note: in this case, you don't have an expression for y in terms of x, so your expression for y' will have both x and y.) Show your work, and check your answer with another group.

> ## Summary Box DT7.1: Implicit Differentiation
>
> Given an equation relating x and y, in which y is implicitly a differentiable function of x, the derivative of y with respect to x can be found by taking the derivative of both sides of the equation with respect to x and then solving for y'. This method is called **implicit differentiation**.

Extend Your Understanding Questions (to do in or out of class)

17. Use implicit differentiation to show that for the last function in Table 2, $x^2 y + y^2 + 7 = 0$, $y' = \dfrac{-2xy}{x^2 + 2y}$.

$2xy' = -2yy'$

$y'(2xy + 2y) = 0$

18. Use implicit differentiation to find y'.

a. $x^3 + 6y^2 = 1$

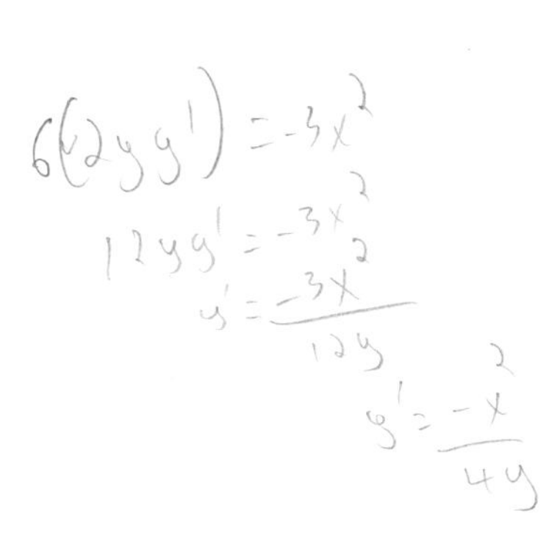

b. $x^2(x+y) = y^3(x-y)$

c. $y\sin^2 x = x\sin y$

d. $\sqrt{xy} = 1 + x^2 y$

e. $\dfrac{x}{7-y^2} = \sin(x-y)$

Notes

Differentiation Applications 1: Related Rates

Model 1: Sliding Ladder

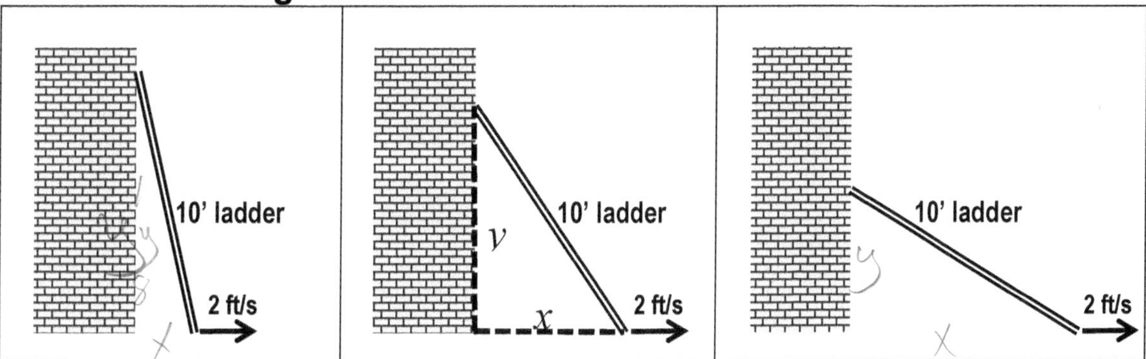

A 10 ft ladder is leaning against a wall when the bottom begins to slip out. Assume that the bottom of the ladder slides away from the wall at a constant rate of 2 ft/s, as represented by the arrow on each drawing. At any moment, the bottom of the ladder is x feet from the wall, and the top of the ladder is y feet from the ground, as shown in the center diagram. Focus on the question:

What is the velocity of the top of the ladder when it is a given distance above the ground?

as you work through the first few pages of this activity.

Construct Your Understanding Questions (to do in class)

1. Label the distances x and y in the first and last drawings in Model 1.
2. When… a. $y = 8$ ft, what is x? b. $y = 3$ ft, what is x?

   Write down any equations you used to find your answers.

   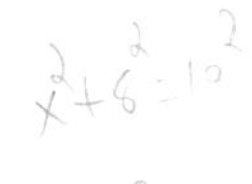

   $3^2+x^2=10^2$

   $x=9.5$

3. Suppose we want to know the velocity of the top of the ladder when $y = 8$ feet. A good way to begin answering this question is to **take an inventory of what is known and unknown, assigning variables to each**. Do this by completing the table below.

Description at the moment of interest: (when y = 8 ft)	**Symbol** or **Variable**	**Value** If unknown, write "?"	**Units**
Distance of the top of the ladder from the ground	y	8	feet
Distance of the bottom of the ladder from the wall	x	6	ft
Velocity of the **bottom** of the ladder, which is the rate of change in x or y [circle one] with respect to time.	$\frac{dx}{dt}$	2	ft/s
Velocity of the **top** of the ladder, which is the rate of change in x or y [circle one] with respect to time.	$\frac{dy}{dt}$	?	ft/s

4. (Check your work) There should be only one unknown in the table in Model 1, and this should correspond to the question (**in bold**) in Model 1. If you listed a value of "?" for both velocities, read Model 1 more carefully and change one of these to an actual value.

5. Note that x and y in Model 1 are changing with time (t), so each is a function of t.

   a. Construct a description of what these symbols mean in the context of Model 1.

      i. $\dfrac{dx}{dt}$ ii. $\dfrac{dy}{dt}$

   b. (Check your work) If you have not already done so, circle x or y in the last two rows on the table in Question 3. Write $\dfrac{dx}{dt}$ and $\dfrac{dy}{dt}$ in the appropriate boxes of the table, and check that your descriptions above match the descriptions in the table.

   c. Is $\dfrac{dy}{dt}$ **positive** or **negative**? Circle one, and explain your reasoning.

   d. (Check your work) Is your answer to part c consistent with the fact that $\dfrac{dx}{dt}$ and $\dfrac{dy}{dt}$ have opposite signs in this problem?

6. (Check your work) Did you cite the equation $x^2 + y^2 = 100$ in Question 2?

   a. If not, go back and check your work.

   b. Starting from this equation, use implicit differentiation to generate an expression for $\dfrac{dy}{dt}$, in terms of the variables and rates of change on the table in Model 1. (If you are stuck, move onto parts c-e on the next page.)

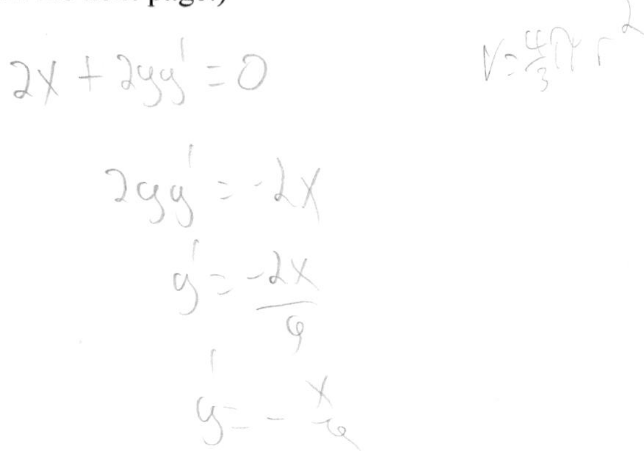

c. (Check your work) Which is the derivative of the term x^2 **with respect to *t***?

$2x$ $\quad 2x\cdot\frac{dy}{dx}$ $\quad 2x\cdot\frac{dx}{dt}$ $\quad \frac{dy}{dx}$ $\quad \frac{dx}{dt}$ $\quad x^2\cdot\frac{dy}{dx}$ $\quad x^2\cdot\frac{dx}{dt}$

Circle one and state what differentiation rule is being applied here.

d. Using this same differentiation rule, find the derivative **with respect to *t*** of the term y^2.

e. Use your answers to parts c and d, above, to construct an answer to part b.

Check the resulting expression for $\frac{dy}{dt}$ with another group.

7. Use what you have learned so far in this activity to find the velocity of the top of the ladder in Model 1…

a. When it is 8 feet above the ground? Draw a picture and show your work.

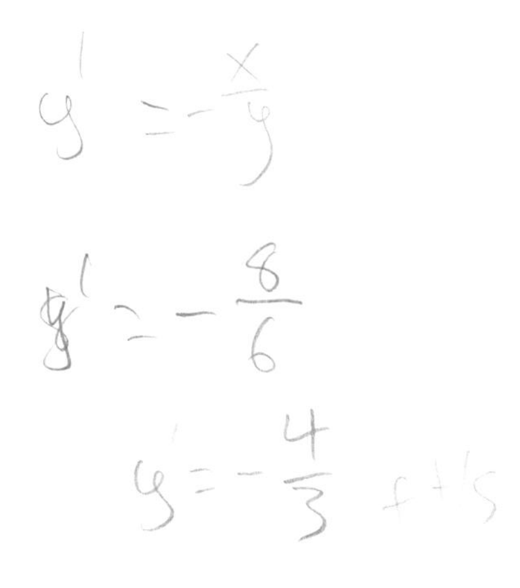

b. When it is 3 feet above the ground? Draw a picture and show your work.

8. Helium is leaking out of a spherical weather balloon at a constant rate of π cubic feet per second. What is the rate of change in the radius of the balloon at the moment when the volume of the balloon is $33.5\ \text{ft}^3$? It is helpful to draw a picture and you may summarize the information in a table like the one in Model 1. If you are stuck, try the next question, then come back to complete this question. Note that the volume of a sphere is given by $V = \frac{4}{3}\pi r^3$

Description at the moment of interest: when…	**Symbol** or **Variable**	**Value** If unknown, write "?"	**Units**
Volume	y	33.5	ft^3
radius	X	2	ft
rate of change of radius	$\frac{dx}{dt}$	?	ft/sec
rate of volume	$\frac{dy}{dt}$	π	ft/sec

$33.5 = \frac{4}{3}\pi r^3$

$25.125 = \pi r^3$

$8 = r^3$

$r = 2$

$\frac{dx}{dt} = \pi$

$\frac{dy}{dt} = \frac{4}{3}\pi r^3$

$\frac{4\pi}{3} 3r^2$

$\frac{12\pi}{3} r^2 \quad 4\pi r^2$

$\frac{dy}{dt} = \pi$

$\frac{dx}{dt} =$

9. (Check your work) The previous question can be solved by answering each of the following questions. If you skipped any of these, answer the question, then go back consider how it applies to the previous question.

   a. What defines the moment of interest for this problem? (e.g., in Model 1, the moment of interest was defined by the time at which the top of the ladder was 8 ft. from the ground, or $y = 8$ ft.)

   Volume = 33.5

   b. What variables and rates of change are involved in this problem? Write a description and assign a symbol and units to each one. (Enter these on the table.)

π ftp/s $- \frac{dx}{dt}$ $\quad \frac{dy}{dt} = 4\pi r^2$

c. Write down an equation relating the variables in the problem.

d. Consider the equation, $y = k \cdot x^3$, where k is a constant and the variables x and y are each a function of time, t.

   i. Use implicit differentiation to generate a new equation containing the rate of change of x with respect to t (and other quantities), then solve for this rate. Show your work.

   ii. Explain how this relates to the weather balloon problem.

e. What is the radius when the balloon volume is 33.5 ft^3?

f. What is the sign of $\frac{dV}{dt}$?

g. What is the rate of change in the radius of the balloon at the moment when the volume of the balloon is 33.5 ft^3 ?

10. (Check your work) The following is a checklist of steps that could be useful in solving a related rates problem. Which steps did you use in solving Question 8? Which steps did you skip? Put a check mark next to the ones you used.

   o Draw a picture and assign symbols to relevant quantities.

   o Summarize the information provided in the problem (e.g., make a table):

      - Note which quantities are changing over time, and use correct notation to express their rate of change with respect to time.
      - Write a description of what each variable and rate of change means, being mindful of units of measure.
      - Identify which quantities are known and which are unknown.

   o Write down an equation relating the variables you identified.

   o If possible, use geometry to eliminate one or more variables.

   o Use implicit differentiation to generate a new equation relating the variables and their rates of change, then solve for the unknown.

   o Answer the question asked in the problem remembering to state the units of measure.

Extend Your Understanding Questions (to do in or out of class)

11. Two straight roads intersect in the middle of town. One runs north-south, the other east-west. Noah is driving north from the intersection at 20 miles per hour. Ellie is driving east from the intersection at 30 miles per hour. When Noah is 0.3 miles from the intersection and Ellie is 0.4 miles from the intersection, how fast are they moving apart?
(Try not to look back at your checklist unless you are stuck.)

12. The conical water tank shown at right is filling at a rate of 1 m^3 (1000 L) per hour.

<u>At what rate is the water level rising when the depth of the water is 2 m</u>?

Recall that, for a cone, $V = \frac{1}{3}\pi r^2 h$.

(Try not to look back at your checklist unless you are stuck.)

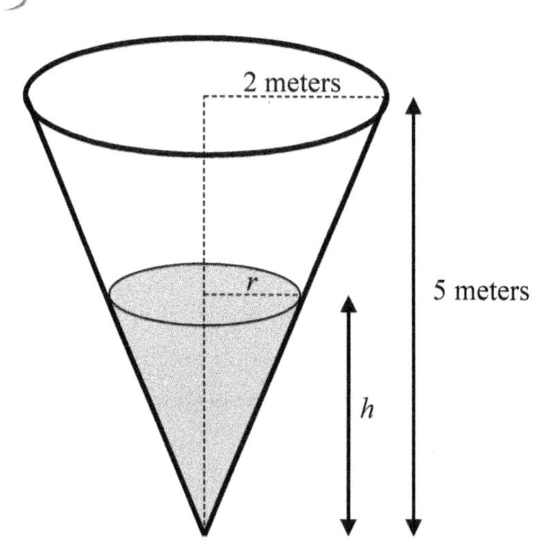

13. (Check your work) Jeff generates the following solution to the previous question:

$\frac{r}{h} = \frac{2}{5}$ (by similar triangles) so $r = \frac{2h}{5}$

At the moment of interest: $h = 2$, so $r = \frac{2(2)}{5} = 0.8$

Substituting $r = 0.8$ into $V = \frac{1}{3}\pi r^2 h$ gives...

$V = \frac{1}{3}\pi(0.8)^2 h \approx 0.67h$

Differentiating $\frac{dV}{dt} = 0.67\frac{dh}{dt}$ and solving gives $\frac{dh}{dt} = 1.5\frac{dV}{dt}$

Plugging in $\frac{dV}{dt} = 1$ m^3/hr gives the answer as $\frac{dh}{dt} = 1.5$ m/hr

a. (Check your work) Jeff did a good job setting up the variables; and his equation $r = \frac{2h}{5}$ is correct. Are these choices consistent with your entries in the table on the previous page? If not, go back and check your work.

b. The correct answer to this question is $\frac{dh}{dt} = \frac{25}{16\pi} \approx 0.5$ m/hr

Jeff made a very common error for this type of problem. He eliminated a variable by substituting in an actual value too soon. Find Jeff's error, correct it, and solve the problem to get the right answer. Show your work.

c. (Check your work) To solve this problem Jeff should have substituted $r = \frac{2h}{5}$ in the volume equation before using implicit differentiation (not the value 0.8). Is this consistent with your answer to part b? If not, go back and revise your work.

Notes

DA 2: Linear Approximation

Model 1: Sleepy Road Trip

You are on a road trip with your family and have been traveling for nearly two hours on small roads when you get on the highway. Soon after entering the highway, you fall asleep. The trip odometer was reset to zero at the start of the trip, and right before you fell asleep you notice it read 60 km, and the speedometer read 60 km/hour (1 km/min).

Construct Your Understanding Questions (to do in class)

1. With only the information in Model 1, make your best estimate of what the trip odometer will read when you wake if you sleep for 60 minutes. Explain your reasoning.

   120

2. $f(t)$ is the distance of the car from the starting point after t minutes. If you fell asleep at $t = 120$ minutes, and Δt is the length of time you were asleep, complete the following table.

Symbol	Units	Value	Description
Δt	minutes	60	length of time you were asleep
$f(120)$	Km	60	distance when fell asleep
$f'(120)$	km/min	speed at 120 min 60	speed at fell asleep
$f(120+\Delta t)$	km	120 Estimate this based on your answer to Question 1 (actual value is unknown)	

3. Use words to fill in the blanks in the following sentence about the Sleepy Road Trip:

   The trip odometer reading at the moment you woke is approximately equal to the trip odometer reading at the moment ______ plus the speed of the car at the moment ______ × ______.

4. Use symbols from the table to turn the sentence in the previous question into an expression for... $f(120+\Delta t) \approx$ $f(120)+f'(120)\Delta t$

5. (Check your work) Show how the expression you generated in the previous question was (or can be) used to answer Question 1. If your expression does not work in Question 1, go back and reexamine your answers to the previous two questions.

6. The graph at right shows what happened during the Sleepy Road Trip in Model 1.

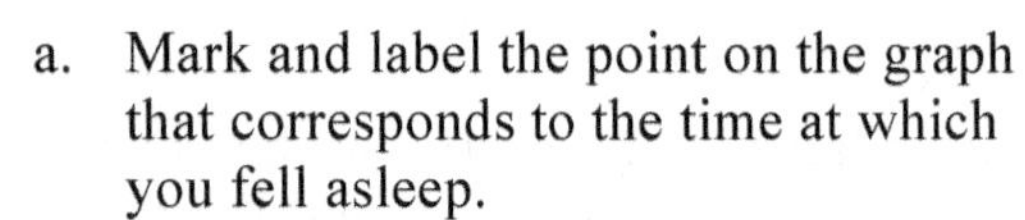

   a. Mark and label the point on the graph that corresponds to the time at which you fell asleep.

   b. (Check your work) Does the point labeled "60 min nap estimate" match your answer to Question 1? If not, go back and reconsider Question 1.

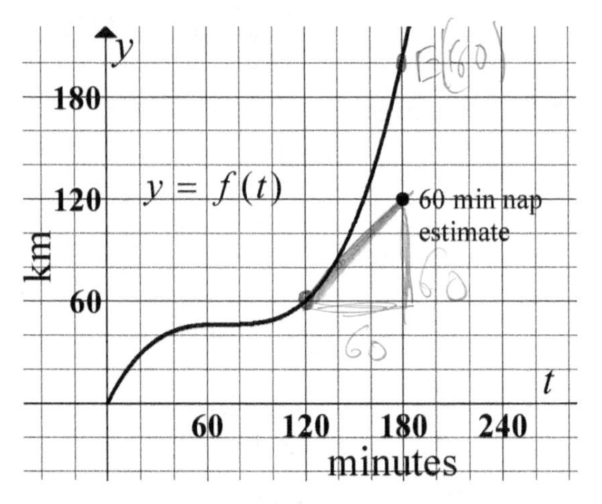

   c. Mark the point on the graph corresponding to the <u>actual</u> odometer reading the moment you woke from your 60 minute nap.

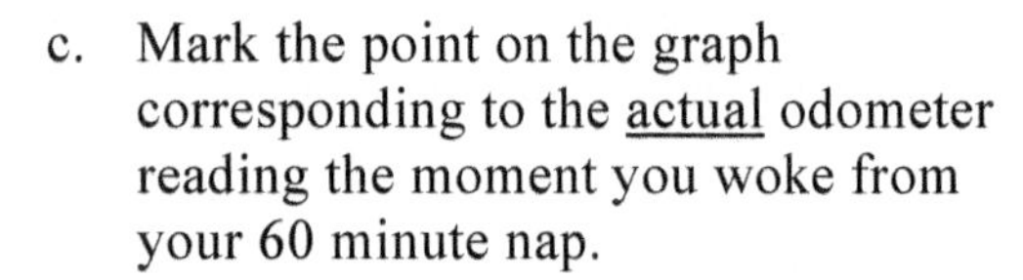

   d. The distance between the actual and estimated odometer readings is the error in this estimate. Mark this vertical gap on the graph and label it $E(180)$ since it is the *E*rror in your estimate at $t = 180$. Based on the graph, the value of $E(180) =$ __60__ km.

   e. Why was the estimate off? What incorrect assumption was built into the estimate you made in Question 1?

   change in speed

7. On the graph above, draw a line from the point you labeled "fell asleep" to the point labeled "60 min nap estimate."

   a. Describe the difference between the imaginary trip represented by this line versus the actual trip represented by the function f during the interval from $t = 120$ to 180.

   b. This line is the tangent line to the graph of $y = f(t)$ at $t =$ __120__ minutes.

   c. Use the graph to find the slope of this tangent line. Show your work.

   d. Which of the following expressions best describes this slope? [circle one]

   $f'(120)$ $120 - f(120)$ $f'(180)$ $f(180) - f(120)$

8. The estimate you made in Question 1 is called a **linear approximation**. Construct an explanation for why this type of approximation is called a *linear* approximation.

9. Use a linear approximation based on the position and speed of the car at $t = 120$ to estimate the position of the car at $t = 140$. (Write down any assumptions and show your work.)

80km

   a. On the graph in Question 6, mark a point that corresponds to your linear approximation of the position of the car at $t = 140$.
   (Check your work) This is *not* the point on the curve at $(140, 85)$.

   b. What is the error in this approximation? $E(140) =$ 5 km

10. Fill in the blanks: By analogy to Model 1, the linear approximation in the previous question corresponds to a nap starting at $t =$ 120 minutes, and lasting 20 minutes.

11. In this activity you used the position and speed of the car in Model 1 at $t = 120$ to estimate the car's position at $t = 140$ and at $t = 180$. Even without the graph of $y = f(t)$ shown in Question 6, one would guess that the estimate at $t = 140$ is more accurate than the estimate at $t = 180$. Explain.

12. Give an example of a value of t for which a linear approximation using the tangent line to f at $t = 120$ on the previous page would give an error smaller than either of the estimates we have made so far (that is, smaller than $E(180)$ or $E(140)$).

125

13. In general, the **smaller** or **larger** [circle one] the value of Δt in $f(t + \Delta t) \approx f(t) + f'(t) \cdot \Delta t$ the smaller the error in corresponding linear approximation. (Check your work) Is this consistent with your answers to the previous two questions? Explain.

Model 2: Differentials

The graph at right shows a function f in bold, and the line tangent to f at P.

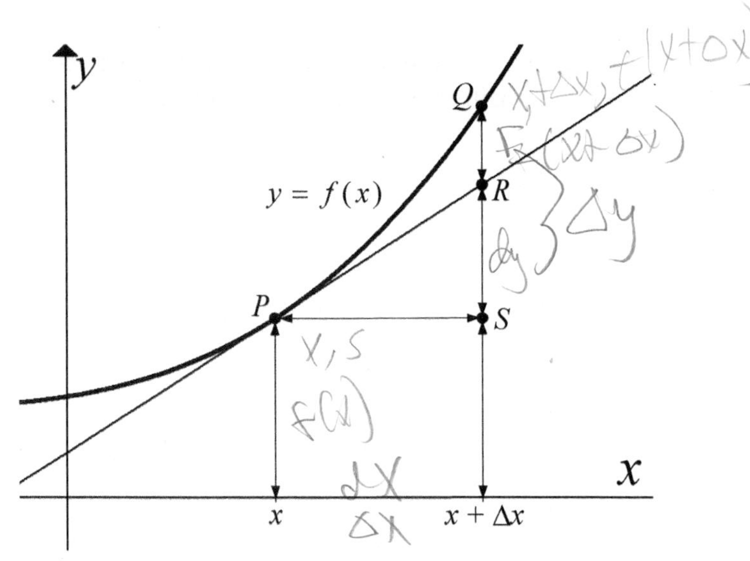

Construct Your Understanding Questions (to do in class)

14. Label the point marked P with its coordinates.

15. Which letter (***Q*** or ***R*** or ***S***) marks the point on the graph with coordinates $(x+\Delta x, f(x+\Delta x))$? Label this point with its coordinates.

16. Label the two double headed arrows on the graph in Model 2 that have length "$f(x)$".

17. It turns out that Δx in Model 2 is sometimes also called dx. Write the label "Δx (or dx)" below the double headed arrow that has a length equal to Δx.

18. Δy is the difference between the value of f at x and the value of f at $(x+\Delta x)$. That is, $\Delta y = f(x+\Delta x) - f(x)$. On the graph, mark the vertical distance Δy with a bracket like this → } and label it Δy.

19. The tangent line in Model 2 is the basis for a linear approximation of the value of f at Q.

   a. What point in Model 2 has a y value equal to this linear approximation value?

   b. $E(x+\Delta x)$ is the error in this linear approximation of the value of f at Q. Write $E(x+\Delta x)$ next to the double headed arrow that represents the size of this error.

   c. For reasons we will explore later, the vertical distance from S to R is called dy. Label this dy on the graph, and mark each of the following statements **True** or **False**.

      i. **T** or **F**: $E(x+\Delta x) = \Delta y - dy$

      ii. **T** or **F**: As Q gets closer to P, $E(x+\Delta x)$ gets smaller

      iii. **T** or **F**: As Q gets closer to P, dy approaches Δy

      iv. **T** or **F**: For Q close to P, $dy \approx \Delta y$

      v. **T** or **F**: $f(x+\Delta x) = f(x) + \Delta y$

      vi. **T** or **F**: $f(x+\Delta x) \approx f(x) + dy$

   d. What is $f'(x)$, the slope of the tangent line in Model 2, in terms of the lengths labeled dy and dx in Model 2? $f'(x) =$ ______

e. (Check your work) Is your answer to part d consistent with the fact that $dy = f'(x) \cdot dx$?

f. Use true statements above to show that $f(x + \Delta x) \approx f(x) + f'(x) \cdot dx$

20. $f(120 + \Delta t) \approx f(120) + f'(120) \cdot \Delta t$ is the correct answer to Question 4. Now we will compare this result to your answer to part f of the previous question:

Let $f(x)$ be a function that gives the position of a car in terms of time, x, in minutes. Describe each of the following symbols in the language of the Sleepy Road Trip in Model 1. (The first one is done for you.)

a. Δx is *the length of the nap* .

b. $f(x + \Delta x)$ is ______

c. $f(x)$ is ______

d. $f'(x)$ is ______

e. dx is ______

f. $f'(x) \cdot dx$ is ______

g. $f(x) + f'(x) \cdot dx$ is ______

______

21. (Check your work) Is your answer to part g, above, consistent with the fact that $f(x + \Delta x)$ is equal to the *approximate* position of the car at the end of the nap based on the linear approximation? If not, go back and check your work.

22. For dy, dx, Δy, and Δx as defined in Model 2 (mark each statement true or false):

a. **True or False:** $dy = \Delta y$ Explain your reasoning.

b. **True or False:** $dx = \Delta x$ Explain your reasoning.

Extend Your Understanding Questions (to do in or out of class)

23. For $f(x) = \sin x$, shown at right.

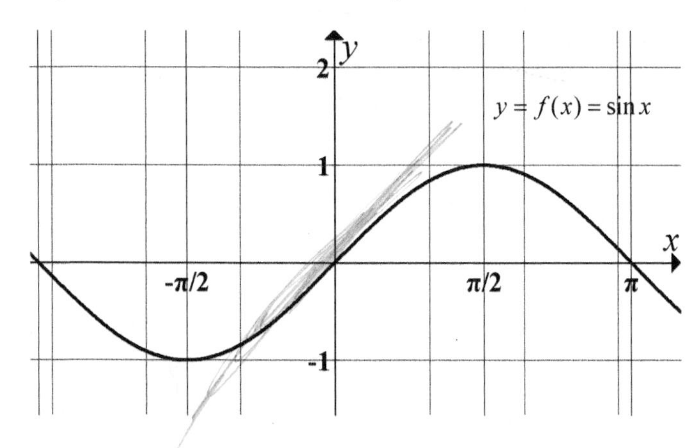

   a. What is $f'(x) =$

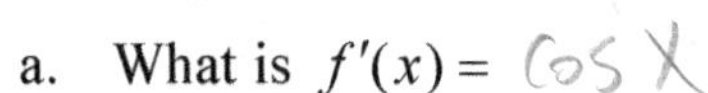

   b. What is $f'(0) =$

   c. On the graph, draw a line tangent to the graph of $y = f(x)$ at $x = 0$.

   d. What is the equation of this tangent line?

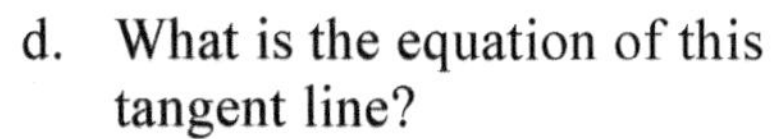

   e. (Check your work) Do your answers to parts b and d of this question agree? If not, revise one or both answers.

24. Use the tangent line you drew above to make a linear approximation of the value of …

   a. $\sin(0.5) \approx$

   b. $\sin(0.1) \approx$

   For each of these, mark a point on the graph in the previous question at the actual value of $f(x)$ and comment on how close this is to the linear approximation value (on the tangent).

25. Which estimate in the previous question is more accurate? Explain your reasoning.

26. Construct an explanation for why using this tangent line to generate an estimate of sin(2) does not make sense. Mark the point at $x = 2$ on both f and the tangent to f that you drew.

27. For the scale drawn on the graph above, cite an interval of x for which it appears the linear approximation at $x = 0$ is indistinguishable or nearly indistinguishable from values of $f(x)$.

28. (Check your work) Do you remember seeing the following in your physics textbook: "$\sin x \approx x$ when x is close to zero"? Is your answer to the previous question consistent with this statement? If not, go back and reconsider your answer.

29. Suppose we want to know the square root of a number near 16 (e.g. 18), but we do not have a calculator. One technique is to use a linear approximation of the function $f(x)=\sqrt{16+x}$ at $x=2$.

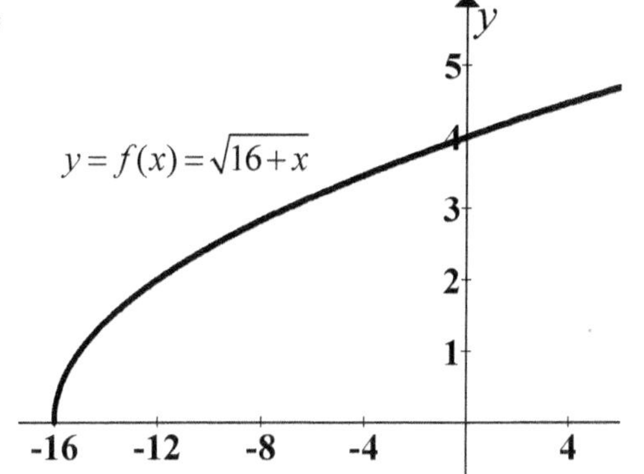

   a. At $x=0$, what is the value of f? That is, $f(0)=$ 4

   b. At what value of x is $f(x)=\sqrt{18}$? 2

   c. For $f(x)=\sqrt{16+x}$, what is $f'(x)=$

   d. What is $f'(0)=$

30. We know that $\sqrt{18}$ is between 4 and 5 since $4^2=16$, and $5^2=25$, but if we want a more *specific* estimate we can use a linear approximation.

   a. Use $f(x)=\sqrt{16+x}$ and the tangent line to the graph of $y=f(x)$ at $x=0$ to make a linear approximation of the value for $\sqrt{18}$. Hint: start by drawing the tangent line on the graph at $x=0$, and finding its slope. Write down all formulas and show all work.

   b. (Check your work) Did you find that the slope of the tangent line in your linear approximation is equal to your answer to part d of the previous question, $f'(0)$?

   c. The tangent line at $x=1$ would give us a more accurate estimate of $\sqrt{18}$. Why didn't we use this tangent line instead? That is, what is the advantage of using the tangent line at $x=0$ for our linear approximation? (Recall that we are assuming we do not have a calculator.)

31. We could have used many different functions to find an approximate value for $\sqrt{18}$. Explain why, for a tangent line at $x = 0$, the function $f(x) = \sqrt{16 + x}$ makes more sense than...

a. $f(x) = \sqrt{17 + x}$ (with an approximation at $x = 1$)

b. $f(x) = \sqrt{9 + x}$ (with an approximation at $x = 9$) or
$f(x) = \sqrt{25 - x}$ (with an approximation at $x = 7$)

32. (Check your work) A student uses his calculator to find $\sqrt{18}$ and gets: $\sqrt{18} = 4.2426$. Use this value to calculate the error in your linear approximation on the previous page. That is, $E(2) =$ Show your work and explain your reasoning.

33. What value do you get if you use $f(x) = \sqrt{16 + x}$, and the tangent line to the graph of $y = f(x)$ at $x = 0$ to generate a linear approximation for $\sqrt{20}$. Show your work.

34. (Check your work) If we take the (rounded) value of $\sqrt{20}$ to be... $\sqrt{20} = 4.472$, what is the error in your linear approximation in the previous question? That is, $E(4) =$

35. Since this is the error in an estimate of $\sqrt{20}$, you might want to call it $E(20)$, but this is not correct.

a. Explain why this error is called $E(4)$.

b. For $f(x) = \sqrt{16 + x}$, and the tangent line to the graph of $y = f(x)$ at $x = 0$, the notation "$E(20)$" would describe the error in the estimate of the square root of what number?

36. Let us assume that we can tolerate $|E(x)| \leq 0.01$ for the linear approximations in the previous question. That is, our linear approximation is considered valid for values of Δx that result in error less than or equal to 0.01. Based on this criteria, ...

    a. was our linear approximation of $\sqrt{18}$ (that is, $\Delta x = 2$) valid?

    b. was our linear approximation of $\sqrt{20}$ (that is, $\Delta x = 4$) valid?

37. The table below shows entries for $f(x) = \sqrt{16 + x}$.

    a. Which column represents linear approximations of the value of $f(x + \Delta x)$? Explain how you can tell.

    b. (Check your work) Do the table's entries in the column you chose for $\sqrt{18}$ for $\sqrt{20}$ match your calculations in Questions 32a and 35, respectively?

    c. What are the allowable maximum and minimum values of Δx found on the table (based on the criteria in the previous question)?

The number we are trying to find the square root of	Square root of the number in Column 1	Δx	$f(0) + f'(0) \cdot \Delta x$	$E(0 + \Delta x)$
12	3.464102	-4	3.5	0.035898
12.5	3.535534	-3.5	3.5625	0.026966
13	3.605551	-3	3.625	0.019448
13.5	3.674235	-2.5	3.6875	0.013265
14	3.741657	-2	3.75	0.008342
14.5	3.807887	-1.5	3.8125	0.004613
15	3.872983	-1	3.875	0.002016
15.5	3.937004	-0.5	3.9375	0.000496
16	4	0	4	0
16.5	4.062019	0.5	4.0625	0.000480
17	4.123106	1	4.125	0.001894
17.5	4.183300	1.5	4.1875	0.004199
18	4.242641	2	4.25	0.007359
18.5	4.301163	2.5	4.3125	0.011337
19	4.358899	3	4.375	0.016101
19.5	4.415880	3.5	4.4375	0.021619
20	4.472136	4	4.5	0.027864

Notes

DA 3: The Mean Value Theorem

Model 1: Pennsylvania Turnpike

You are traveling east on the Pennsylvania Turnpike. You note the time as you pass the Lebanon/Lancaster Exit (A). One hour later you pass the Willow Grove Exit (B). Questions on this page refer to the one hour segment during which you traveled the 75 miles from A to B.

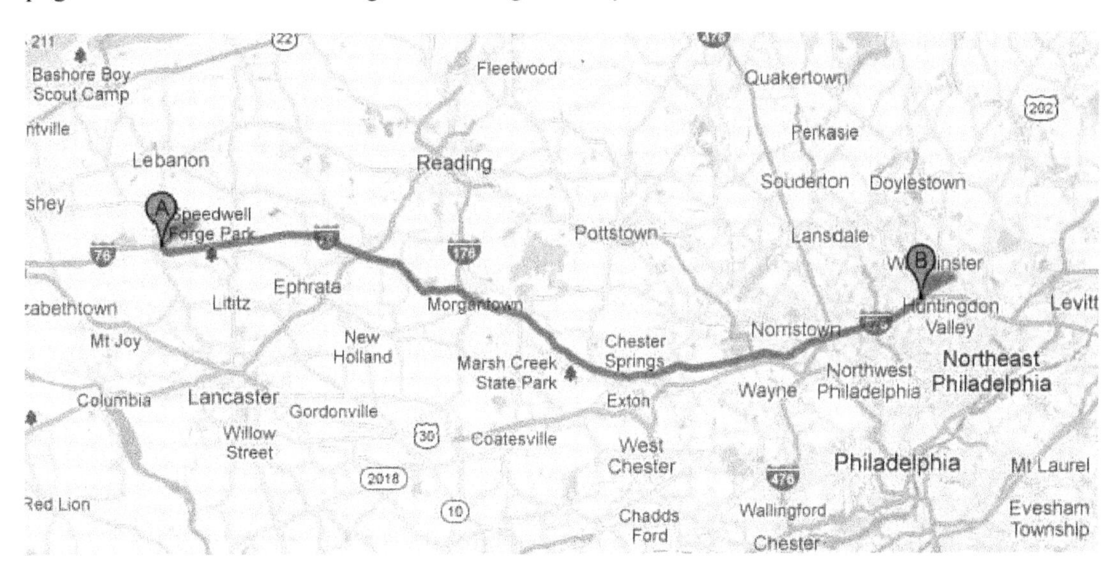

Construct Your Understanding Questions (to do in class)

1. **True** or **False**: It must be the case that you were going 75 miles per hour (mph) during the entire one-hour segment described in Model 1. Circle one and explain your reasoning.

2. You remember looking at your speedometer somewhere in the middle of the segment and your speed was 80 mph. **True** or **False**: This means that at some point during your trip you must have been going less than 75 mph. Circle one and explain your reasoning.

3. **True** or **False**: It must be the case that at some point during this 1-hour segment you were going exactly 75 mph. Circle one and explain your reasoning.

4. **True** or **False**: For any car trip, there is a moment when your instantaneous velocity is equal to the average velocity for the trip. Circle one and explain your reasoning.

5. On the graph of $y=f(x)$ at right...

   a. Draw the secant line through **A** and **B** (the straight line from **A** to **B**).

   b. Mark the point **C**, with coordinates $(c,f(c))$, such that the line tangent to f at **C** has the same slope as the secant line through **A** and **B**.

   c. (Check your work) Draw the tangent line to f at **C**, and confirm that this line is parallel to the secant line through **A** and **B**.

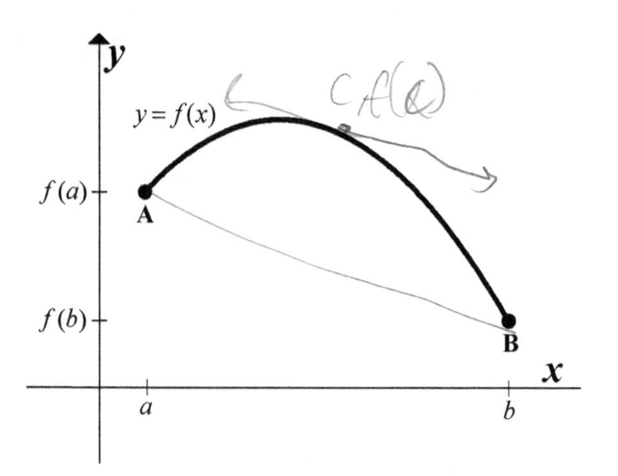

6. Which of the following describe the slope of the straight lines you drew in the previous question? More than one answer is correct. Circle all that apply.

$f(c)$ $\quad f'(a)$ $\quad f'(b)$ $\quad f'(c)$ $\quad f(b)-f(a)$ $\quad \dfrac{f(b)-f(a)}{b-a}$ $\quad \dfrac{f(b)-f(a)}{f(c)}$

7. For each graph, mark one or more points **C**, with coordinates $(c,f(c))$, in the interval (a,b) where $f'(c)=\dfrac{f(b)-f(a)}{b-a}$.

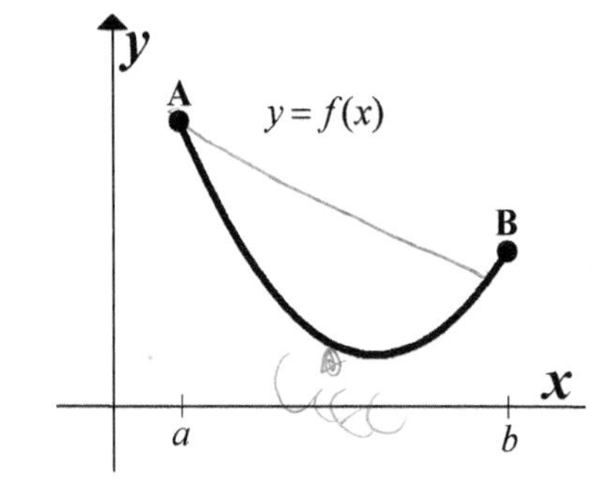

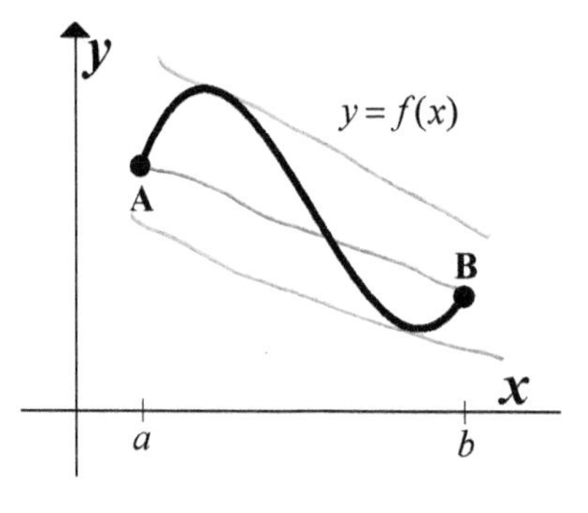

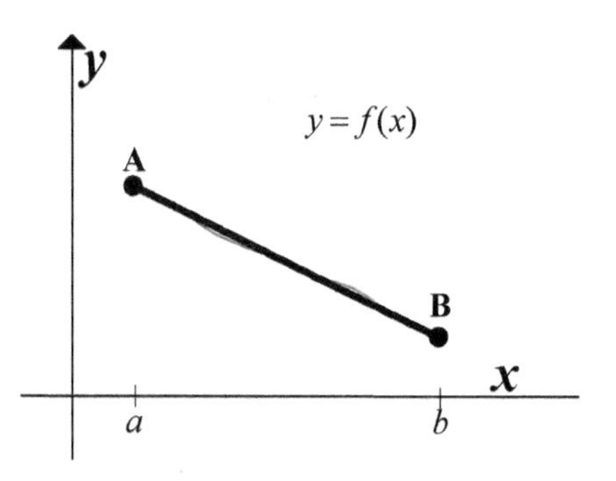

8. On the axes at right, try to draw a function that starts at **A** and ends at **B** but has no point **C** in (a,b) where $f'(c)=\dfrac{f(b)-f(a)}{b-a}$.

   Again assume **C** has coordinates $(c,f(c))$.

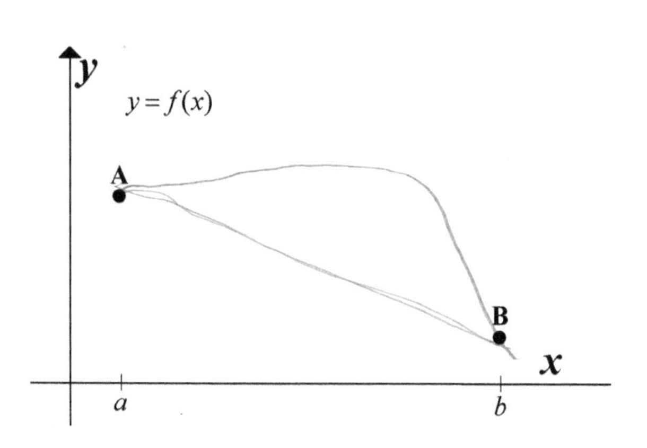

9. Is your answer to the previous question consistent with Summary Box DA3.1? Explain.

Summary Box DA3.1: Mean Value Theorem (MVT)

If f is continuous on $[a,b]$ and differentiable on (a,b) then...

- there exists a number c in (a,b) where $f'(c) = \dfrac{f(b)-f(a)}{b-a}$

  or, in other words...

- there is a point $(c, f(c))$ such that the slope of the line tangent to f at c is equal to the slope of the secant line between $(a, f(a))$ and $(b, f(b))$.

10. For each graph, try to find a point **C**, with coordinates $(c, f(c))$, in the interval (a,b) where $f'(c) = \dfrac{f(b)-f(a)}{b-a}$.

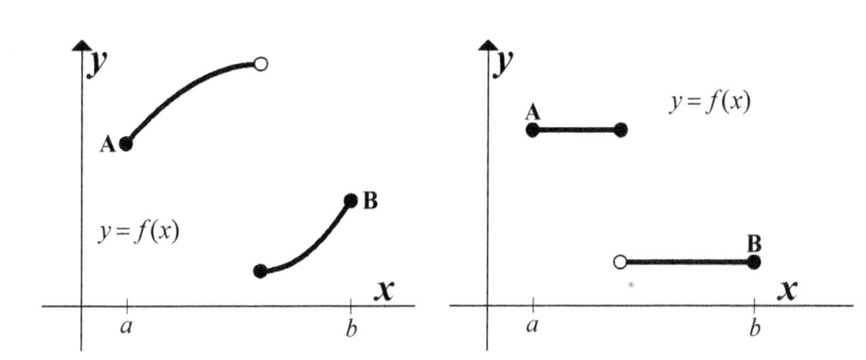

11. Explain why the **Mean Value Theorem** does not apply to the functions in the previous question. That is, explain why there is no point **C** in (a,b) such that $f'(c) = \dfrac{f(b)-f(a)}{b-a}$.

12. (Check your work) In the previous question you likely noted that the functions in Question 10 are not continuous. It turns out that it *is* possible to draw a *continuous* function with no point $(c, f(c))$ in (a,b) such that $f'(c) = \dfrac{f(b)-f(a)}{b-a}$.

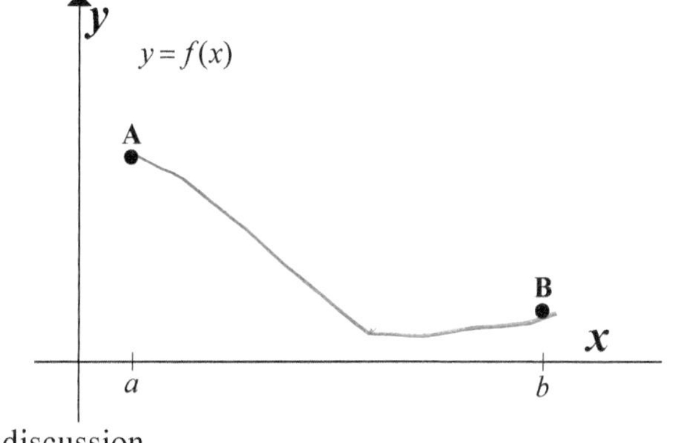

a. Try to draw such a function on the axes at right. Be prepared to share your answer as part of a whole-class discussion.

b. Explain why the **Mean Value Theorem** does not apply to the function you drew.

13. **True** or **False**: For a function f that is continuous *and* differentiable on $[a,b]$, if $f(a)=f(b)$ then there exists a number c in (a,b) where $f'(c)=0$. **If you circled...**

"**True**", draw such a function and mark a point **C** where $f'(c)=0$.

"**False**", draw such a function with no point where $f'(c)=0$.

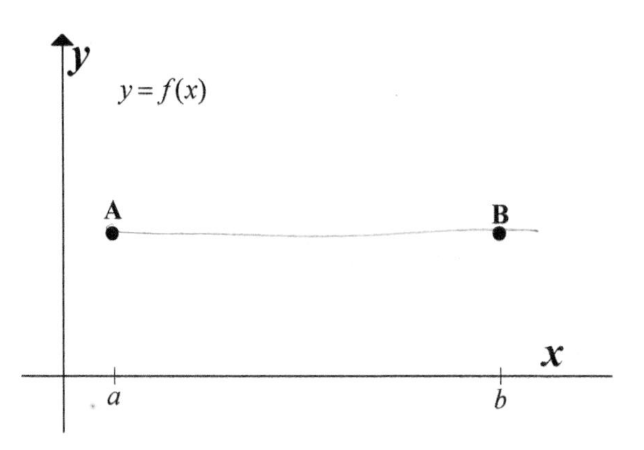

14. *The following question refers to the map of the Pennsylvania Turnpike in Model 1 showing that the distance from the Lebanon/Lancaster Exit to the Willow Grove Exit is 75 miles.*

Ken enters the Pennsylvania Turnpike at **A**. The time-stamped card he receives at the Lebanon/Lancaster entrance to the turnpike says 9:00 PM.

At 10:00 PM he is driving the speed limit (65 mph) when he passes a police officer who is parked near the Willow Grove Exit (**B**). The police officer pulls Ken over and informs him that one of his rear lights is not working. She asks to see Ken's time-stamped card, and then tells him she is going to issue him a ticket for speeding.

Ken exclaims "But officer, I was going the speed limit when I passed you."

Explain how the Mean Value Theorem (MVT) could be used to support the officer's case.

15. Let the functions f and g describe the positions of two horses (*f*rank and **g**eorge) during a race from time $t=a$ to $t=b$. Assume $f'(t)<g'(t)$ for all t in $[a,b]$.

 a. Describe in words what the relation $f'(t)<g'(t)$ tells you about the relative speed of *f*rank and george during the period from time $t=a$ to $t=b$.

b. At right, sketch possible functions f and g showing ***f***rank and ***g***eorge tied at $t = a$.

That is, ***f***rank and ***g***eorge have the same position at $t = a$, as indicated by the point on the graph.

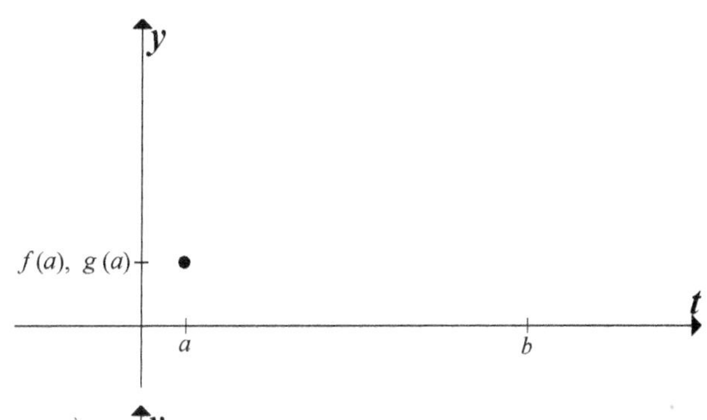

c. At right, sketch a possible pair of functions f and g showing ***f***rank and ***g***eorge tied at $t = b$.

That is, $f(b) = g(b)$.

(Hint: Do not assume this is a fair race and that ***f***rank and ***g***eorge started at same place.)

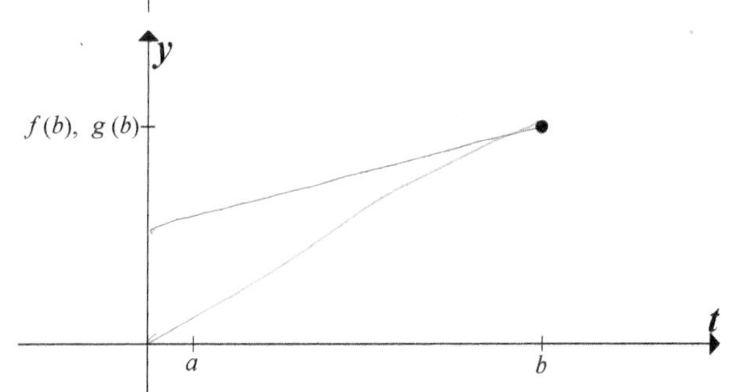

16. (Check your work) Are your answers to the previous question consistent with the fact that (when they are not tied) ***f***rank leads one race, while ***g***eorge leads the other. If not, go back and check your work.

17. The statement in Question 13 is a corollary of the MVT called Rolle's Theorem. Write this label next to Question 13. Other corollaries are given in Summary Box DA3.2. Fill in the blanks to make each true. **Choose from**: increasing, decreasing, constant, $\geq$, $\leq$, or $=$

Summary Box DA3.2: Corollaries of the Mean Value Theorem

If f and g are continuous on $[a,b]$ and differentiable on (a,b) then...

- If $f'(x) > 0$ for all x in (a,b) then f is ___increasing___ on $[a,b]$.
- If $f'(x) = 0$ for all x in (a,b) then f is ___constant___ on $[a,b]$.
- If $f'(x) \leq g'(x)$ for all x in (a,b) and $f(a) = g(a)$, then $f(x)$ ___<___ $g(x)$ for all x in $[a,b]$.
- If $f'(x) \leq g'(x)$ for all x in (a,b) and $f(b) = g(b)$, then $f(x)$ ___<___ $g(x)$ for all x in $[a,b]$.

18. (Check your work) The names of the corollaries in Summary Box DA3.2 are: The Constant Function Theorem, the Racetrack Theorem (tied at the start), the Racetrack Theorem (tied at the finish), and the Increasing Function Theorem. **Match each name to the correct theorem**, and check that your answers are consistent with the graphs you drew in Question 15.

Notes

DA 4: Maximum and Minimum Values

Model 1: Stage 11 of 2012 Tour de France

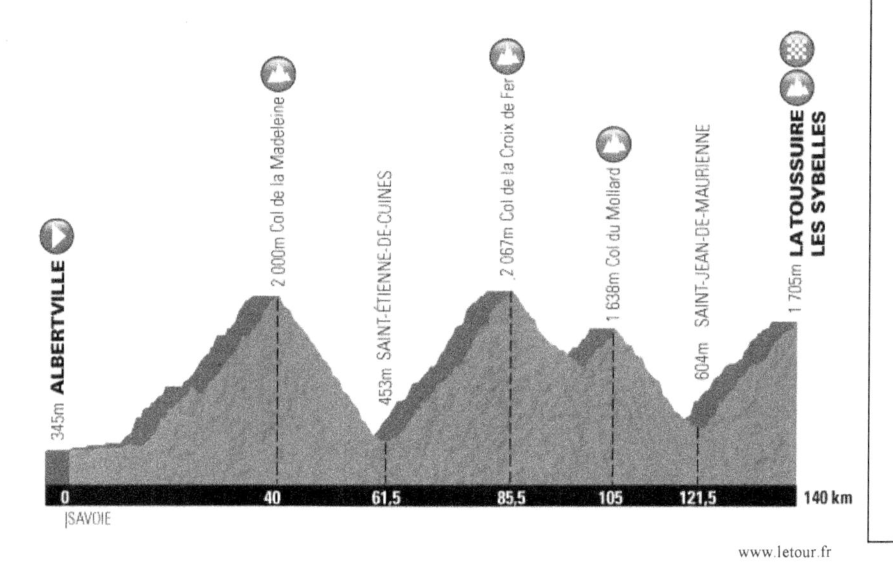

Summary Box DA4.1

For a function f on a domain D, there is a…

global maximum at c if $f(c) \geq f(x)$ for all x in D

global minimum at c if $f(c) \leq f(x)$ for all x in D

local maximum at c if $f(c) \geq f(x)$ for x near c

local minimum at c if $f(c) \leq f(x)$ for x near c

$x =$ distance traveled in km; $f(x) =$ elevation in meters (m). For example, $f(40) = 2000$

Construct Your Understanding Questions (to do in class)

1. Where on the Stage 11 course (at what value of x) is the…

   a. absolute (also called global) maximum value of $f(x)$ found?

   b. absolute (also called global) minimum value of $f(x)$ found?

2. Complete Row 2. In Row 3 write **local max** or **local min** where appropriate, then circle the values of f that correspond to the **global max** and the **global min.**

x	0	40	45	61.5	85.5	105	121.5	138	140
$f(x)$	345	2000	1920	453				1665	
$f(x)$ is a…									

3. On the graph in Model 1…

   a. Use a ▼ to mark one local minimum that is not listed in the previous question.

   b. Use a ▲ to mark one local maximum that is not listed in the previous question.

4. $g(x)$ is continuous on an interval that contains a number c (which is not an endpoint).

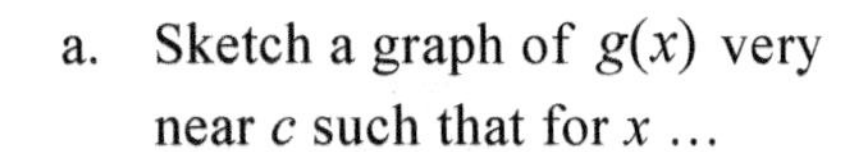

a. Sketch a graph of $g(x)$ very near c such that for x ...

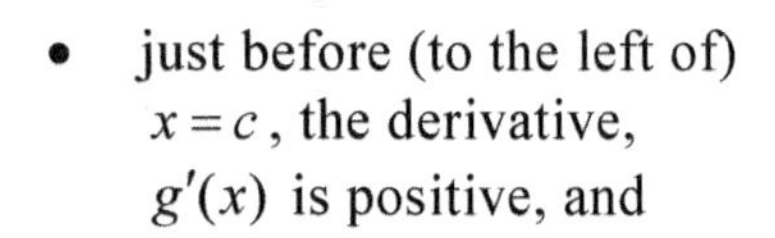

- just before (to the left of) $x = c$, the derivative, $g'(x)$ is positive, and
- just after (to the right of) $x = c$, the derivative, $g'(x)$ is negative.

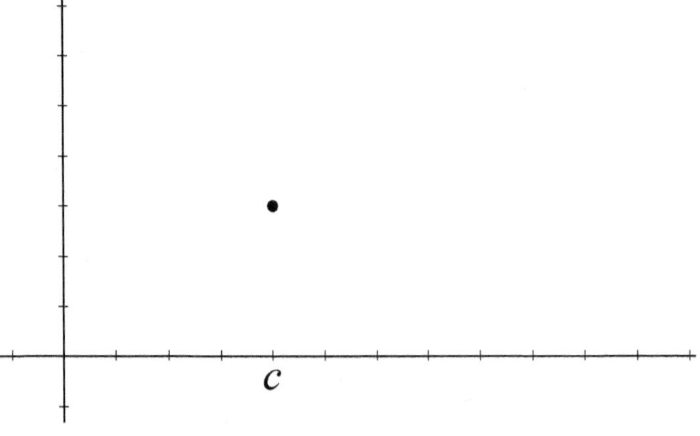

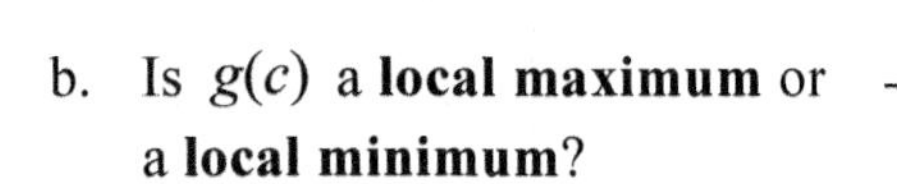

b. Is $g(c)$ a **local maximum** or a **local minimum**?

c. What is the value of $g'(c)$ in your graph? Explain your reasoning.

5. $h(x)$ is continuous on an interval that contains a number c (which is not an endpoint).

a. Sketch a graph of $h(x)$ very near c based on the information that $h'(x)$ is negative just before c and positive just after c.

b. What can you say about $h(c)$?

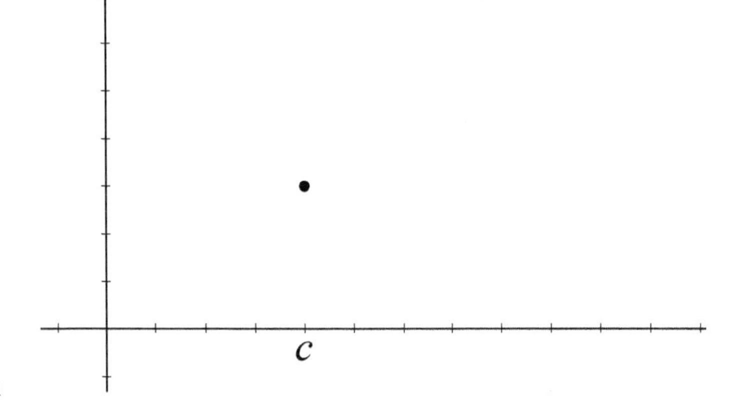

c. What is the value of $h'(c)$ in your graph?

Model 2: Finding Maxima and Minima (Extrema)

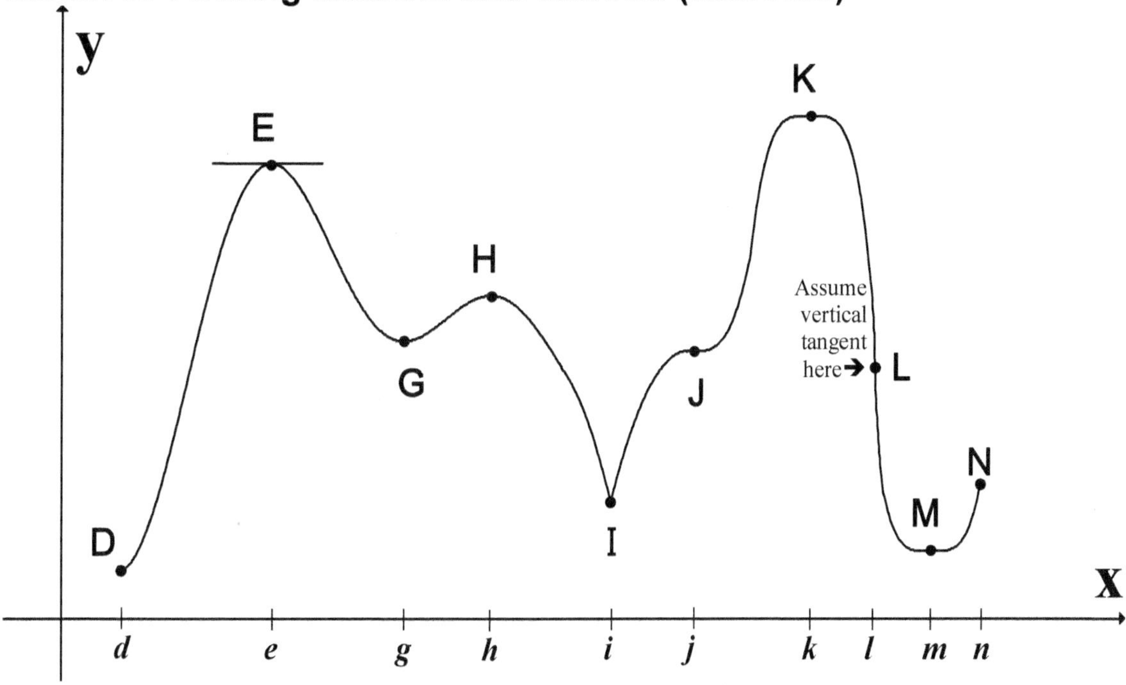

Assume the function $p(x)$ shown in Model 2 is continuous on the closed interval $[d,n]$.

Construct Your Understanding Questions (to do in class)

6. In Model 2, mark each local minimum with a ⮟, and each local maximum with a ⮝.
7. Sketch a short tangent line at the six points in Model 2 where $p'(x)=0$. One is done for you.
8. **True** or **False**: Each point where $p'(x)=0$ corresponds to either a local maximum or local minimum. If false, give an example from Model 2.

9. Write "**DNE**" next to the two points in Model 2 where $p'(x)$ **D**oes **N**ot **E**xist.
10. **True** or **False**: Each point where $p'(x)$ does not exist corresponds to either a maximum or minimum. If false, give an example from Model 2.

11. Consider the **endpoints** of the interval in Model 2, do these endpoints correspond to local extrema? Explain your reasoning.

12. To find all extrema (local and global) for a function f you need only check three categories of points. Two are listed below. What is the third?

    i. endpoints ii. points where $f'(x)=0$ iii.____________________

13. (Check your work) Is your answer to the previous question consistent with Summary Box DA4.2? If not, go back and correct your work.

Summary Box DA4.2: Critical Number

A number c is a **critical number** of f if $f'(c)=0$ or $f'(c)$ does not exist.

14. Complete the following table for selected critical numbers from Model 2.

x	$f'(x)$	$f(x)$ local max, local min, or neither	Sign of $f'(x)$ just before critical number	Sign of $f'(x)$ just after critical number
e	0	local max	+	−
g	0			
h	0			
i	DNE			
j	0			
l	DNE			

15. Based on your answer to the previous question, propose a method to check if a critical number corresponds to a maximum, minimum, or neither.

16. Lola decides to find all local and global extrema for the function $f(x)=x^3+3x^2-24x$ on the interval $[-5,6]$. Correct her work, below. (Her answer is not complete. What did she forget?)

$f(x)=x^3+3x^2-24x$

$f'(x)=3x^2+6x-24$

$f'(x)=3x^2+6x-24=0$

$f'(x)=3(x^2+2x-8)=0$

$f'(x)=3(x+4)(x-2)=0$

$f'(x)=0$ at $x=-4$ and $x=2$

$f'(x)$ exists for all values of x, so the only **critical numbers** of f are $x=-4$ and $x=2$.

$f'(-4.1)=1.8$, $f'(-3.9)=-1.8$ so $f(-4)=80$ is a **local max**

$f'(1.9)=-1.8$, $f'(2.1)=1.8$ so $f(2)=-28$ is a **local min**

Since these are the only local max/min, they are also the global max/min, so $f(-4)=80$ is the **global max**, and $f(2)=-28$ is the **global min**.

17. (Check your work) Is your answer to the previous question consistent with this graph of $f(x) = x^3 + 3x^2 - 24x$ on the interval $[-5,6]$? If not, go back and revise your work.

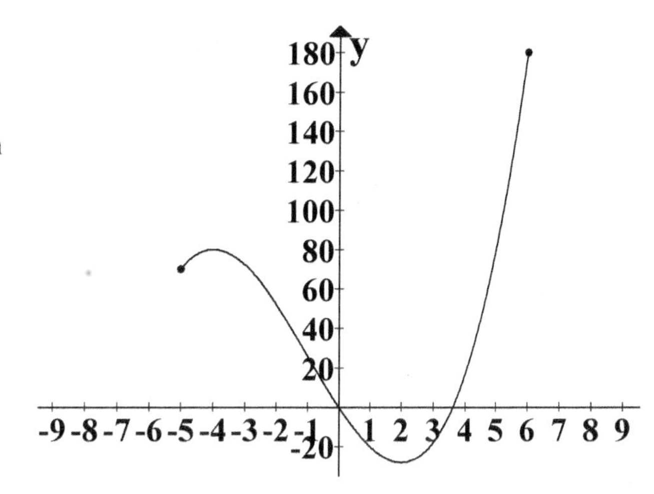

18. Find all local and global extrema for the function $f(x) = 2x^3 + 6x^2 - 90x + 50$ on the interval $[-9,5]$. Show your work.

Model 3: Some Functions Have No Maximum or Minimum Value

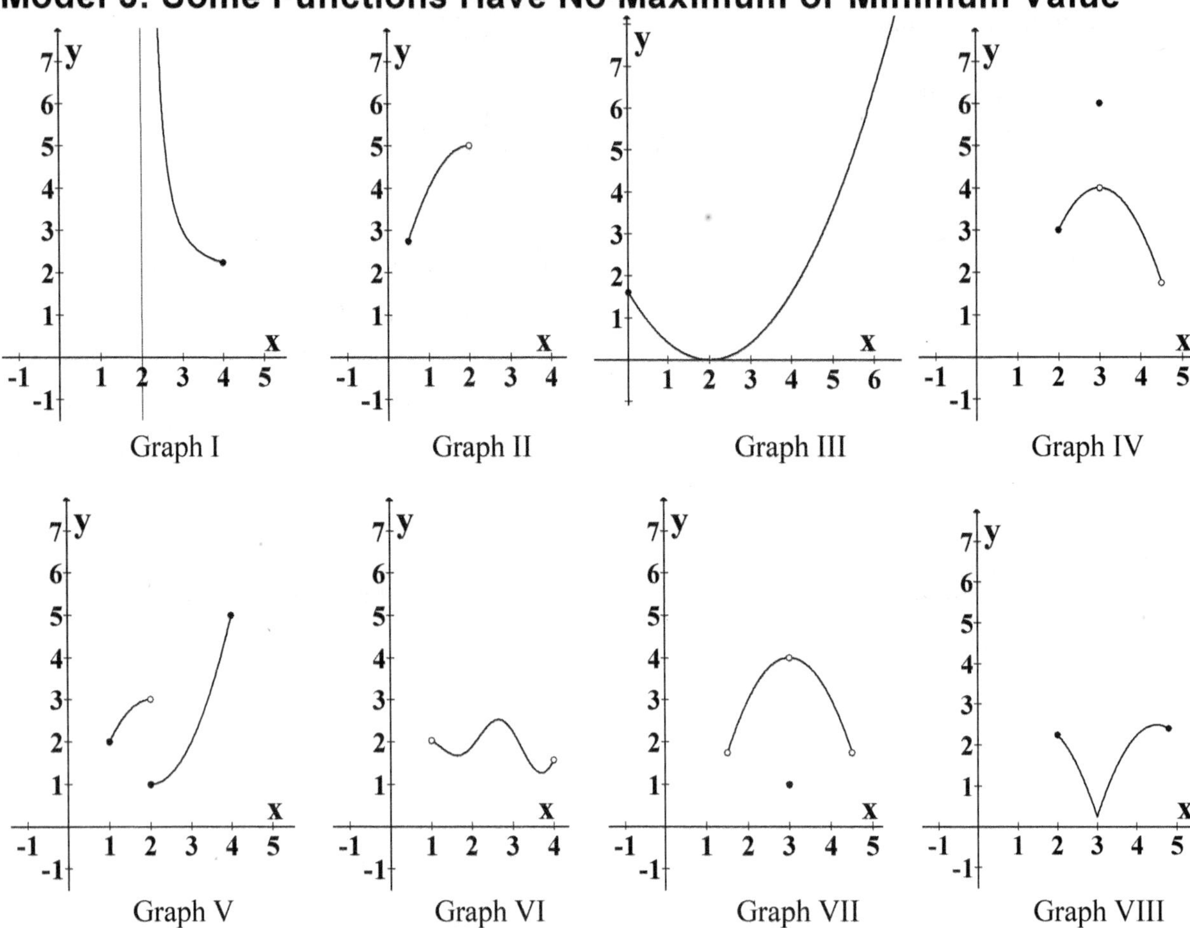

Extend Your Understanding Questions (to do in or out of class)

19. Construct an explanation for why the function shown in Graph I in Model 3 does not have a local or absolute maximum value.

20. Construct an explanation for why the function shown in Graph II in Model 3 also does not have a local or absolute maximum value.

21. Construct an explanation for why the function shown in Graph III in Model 3 has a local maximum value (mark this point "LMax"), but no *absolute* maximum value.

22. Label each absolute maximum (AMax) or absolute minimum (AMin) in Model 3.

    a. Which graph(s) in Model 2 (other than I-III) have no absolute maximum value?

    b. Which graph(s) in Model 2 have no absolute minimum value?

23. Label each local maximum (LMax) and local minimum (LMin) in Model 3 that is not already labeled as an absolute maximum or minimum.

24. Mark each of the following statements as **True** or **False**. For each statement you mark as true, sketch a graph of a function for which it is true. For each you mark as false, sketch a graph of a function that demonstrates that it is false, and explain your reasoning.

    a. If a function has a local maximum then it must have an absolute maximum.

    b. If f is a continuous function on the closed interval $[a,b]$, then f must have an absolute minimum value and an absolute maximum value.

    c. If f has a local maximum or minimum at c then c must be a critical number of f or an endpoint.

    d. For a function f on the open interval (a,b), if f has a local maximum or minimum at c, and $f'(c)$ exists then $f'(c) = 0$.

    e. Every function must have a maximum or a minimum.

    f. Every global maximum (or minimum) is also a local maximum (or minimum).

25. Statement b) in the previous question is called the Extreme Value Theorem (EVT).

    a. Identify the one graph in Model 3 showing a function to which the EVT applies, and explain your reasoning.

    b. Sketch a graph of a function on the closed interval $[a,b]$ that does not have a global maximum or minimum.

    c. Explain why your graph in b) of this question does not satisfy the EVT.

26. For each function, find all local and absolute extrema on the given interval. If there are no extrema of a given type, note this.

    a. $f(x) = 2x+1, \quad x \in [-10,10]$

    b. $f(x) = x^3 + 1, \quad x \in [-1,3]$

    c. $f(x) = \begin{cases} -x, & -1 \le x < 0 \\ x, & 0 \le x \le 3 \end{cases}$

    d. $f(x) = x^3 - 2x^2 - 3, \quad x \in [-2,3]$

    e. $f(x) = x^3 + x^2 - x, \quad x \in [-1.5,1.5]$

    f. $f(x) = x - \sin 2x, \quad x \in [-2,2]$

    g. $f(x) = \ln x, \quad x \in (0,100]$

27. (Check your work) Use a graphing program or calculator to graph each function in the previous question to check your answers.

28. How do your answers for a-g change if the domain of each function changes to an open interval? For example, when part a becomes $f(x) = 2x+1, \quad x \in (-10,10)$.

Notes

DA6: Optimization

Model 1: Max and Minnie Go Camping

Max and Minnie arrive at the campground, a large open field. Max heads to registration, where he is given 120 feet of yellow caution tape and told to mark off a rectangular camp site.

He decides he wants the biggest possible campsite, and (drawing stares from other campers) he exclaims, "My first opportunity to use calculus and it's not even noon!" In his notebook he makes the following diagram of the campsite using x and y to represent its unknown dimensions.

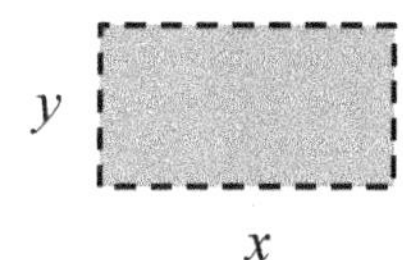

Construct Your Understanding Questions (to do in class)

1. Help Max devise an equation for each of the following **in terms of x and y**.

   a. The area of their campsite: $A =$

   b. The length of the yellow caution tape: $L = 120$ ft. $=$

2. One of the equations in Question 1 introduces a constraint. Without this the maximum area of the campsite could be infinite. Decide which is the **constraint equation** and explain your reasoning.

3. Use both equations in Question 1 to generate a new equation for the area of the campsite in terms of x **only**. This will be a function, $A(x)$. Show your work.

   $A(x) =$

4. The function on the graph at right can be used to model the area of Max's campsite.

   a. Mark the value of x where the area of the campsite is maximized.

   b. **True** or **False**: The value of x you marked is a critical number of $A(x)$. Explain your reasoning.

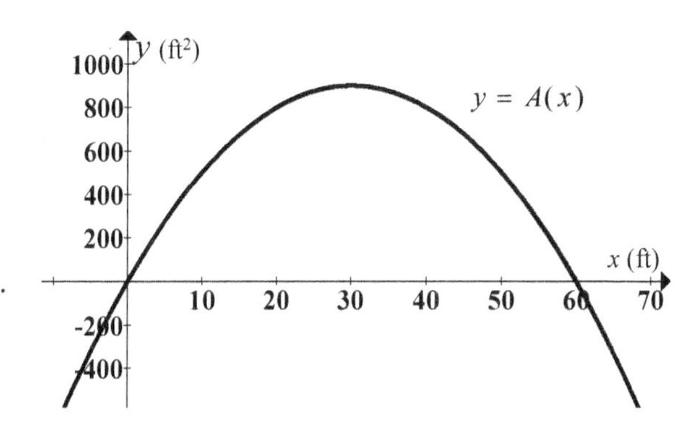

   c. $y = A(x)$ is a parabola whose domain is all real numbers; however, in the context of the area of the campsite, only some values of x have meaning. On the graph, mark the endpoints of the interval where x has meaning, and explain why values of x outside this interval do not make physical sense for this particular problem.

5. Assume you have the formula, $A(x) = -x^2 + 60x$, but you do not have the graph of $A(x)$.

   a. Show that $x = 30$ is a critical number of $A(x) = -x^2 + 60x$.

   b. Show that the global maximum does not occur at either endpoint ($x = 0$ or $x = 60$).

   c. How do you know that $f(30)$ is the global maximum?

6. (Check your work) Max has nearly found his answer when he notices Minnie looking over his shoulder. She says, "Oh, Max. You always do things the hard way. You don't need calculus to maximize the size of our campsite. Like *you,* it's a *square!*" Is Minnie's solution consistent with your answer to the previous question?

7. When finding the maximum (or minimum) of a function on a closed interval, you must check the y values corresponding to the critical numbers ***and the endpoints***. Sketch a graph of a function on a closed interval for which...

   a. the maximum value on that interval is **not** found at a critical number of the function.

   b. the minimum value on that interval is **not** found at a critical number of the function.

8. Max and Minnie walk to the camping area and realize they can make a much bigger campsite if they use an existing fence as one of their boundaries, as in the diagram at right. Again, the area of the campsite can be expressed as $A = xy$.

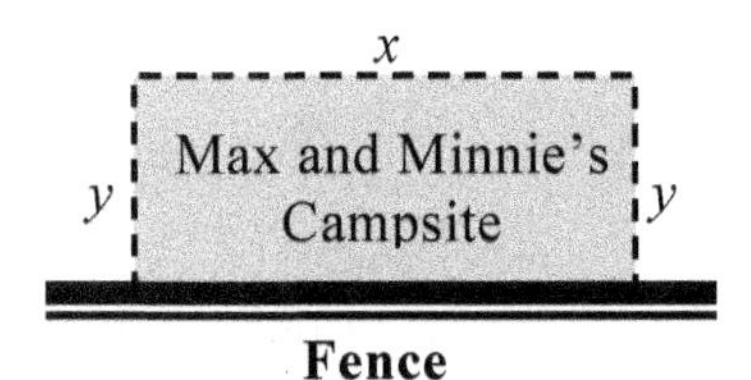

   a. Write a constraint equation analogous to the one you generated in Question 1b.

   b. Use $A = xy$ and your constraint equation to generate a function $A(x)$ that describes the area of the campsite in terms of x only. Show your work.

c. Find the dimensions of the largest possible campsite if Max and Minnie use the fence as one boundary, as shown on the previous page. (If you are stuck, read the next question first.)

9. (Check your work) In the previous question…

   a. Did you find the value(s) of x where $A'(x) = 0$? Explain why this is a *critical* step in finding the solution.

   b. Over what interval does x have meaning? (What are the endpoints of this interval?)

   c. Did you check if these endpoints provide a better solution than the critical number(s)? If not, check these endpoints and show your work here.

10. In each box is a step in the process of solving a max/min problem like the one in Question 8. Number these steps in order (1-8) to generate a useful checklist. The first one is done for you.

Evaluate the y-values of the critical points and endpoints by plugging them into the function being optimized. The largest y-value is the global maximum, and the smallest y-value is the global minimum.	Use the constraint equation to write a new formula for the quantity being optimized that is a function of one variable.
Assign variables to quantities that change.	**1.** Make a drawing.
Find the derivative and then the critical points of the function being optimized.	Identify the constraint equation.
Identify the endpoints, that is, the domain of the function being optimized.	Identify and write down a formula for the quantity that is being optimized.

Model 2: A Better Water Bottle

Minnie is looking at her 40 cm high, 1 liter (1000 cm^3) water bottle. (Assume all water bottles in this activity are perfect cylinders of height h and radius r.) She says to Max, "I think my water bottle is too tall and skinny. If it were shorter and wider it would have a smaller surface area and therefore absorb heat more slowly." Max replies that he thinks all 1 liter cylinders have the same surface area. Minnie decides to investigate by making a table showing her tall bottle (T) and a shorter bottle (S).

Bottle	h (cm)	r (cm)	V (cm^3)	A (cm^2)
T	40	$\frac{5}{\sqrt{\pi}}$	1000	$50 + 400\sqrt{\pi}$ (≈ 758.98)
S	10	$\frac{10}{\sqrt{\pi}}$	1000	

Construct Your Understanding Questions (to do in class)

11. Which formula in the box below represents the volume V of a cylinder, and which represents its outside surface area, A = area of the 3 pieces (below, right) labeled sides, top, bottom.

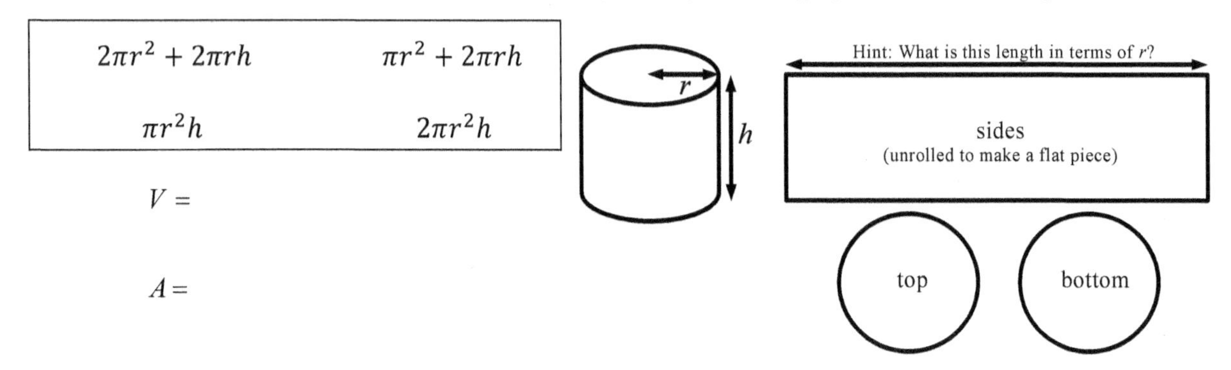

12. (Check your work) Use the formulas you chose for V and A to...

    a. confirm V and A listed for cylinder "T" b. calculate A for cylinder "S"

13. **True** or **False**... a. Max is right: All 1 liter cylinders have the same surface area.

    True or **False**... b. Minnie is right: There exists a shorter, wider 1 liter cylinder with a smaller surface area than her bottle "T".

14. Minnie decides to find the radius of the 1 L cylinder with minimum surface area.
    a. What variable does she need to maximize or minimize? Identify an equation for this variable in terms of h and r.

    b. Identify the constraint equation, and solve this equation for h in terms of r.

    c. Use the constraint equation to rewrite the equation for area in terms of r only.

    d. Describe in words what the function in part c tells you. (Hint: This function does not apply to all cylinders, it only applies to certain cylinders. Which ones?)

    e. Which steps in your checklist in Question 10 have you completed so far?

    f. Use other relevant items from the checklist to solve this problem. Show your work.

Extend Your Understanding Questions (to do in or out of class)

15. The function shown at right models the surface area of a 1 liter cylinder as a function of its radius.

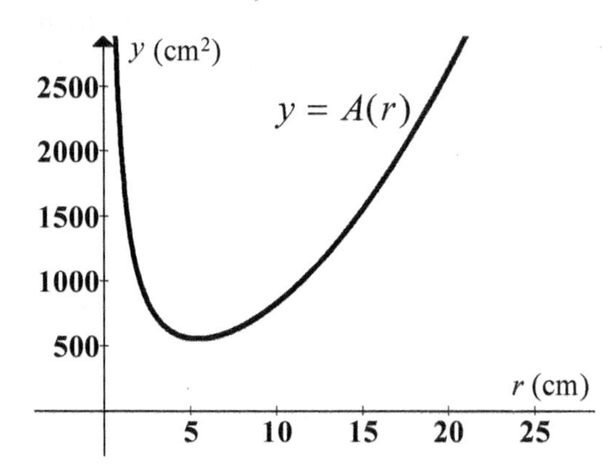

   a. (Check your work) Is this graph consistent with your answer to the previous question? Explain how you can tell.

   b. (Check your work) If you answer is not consistent with the graph, or you did not get an answer, then the following may help determine if you were on the right track.

      i. Did you find that the height, in terms of the radius, of a 1 liter cylinder is $h = \dfrac{1000}{\pi r^2}$, or equivalent?

      ii. Did you use this to substitute for h in your equation for A, and generate a function $A(r)$, that gives the surface area of a 1 L cylinder in terms of r only?

      iii. Did you take the derivative of $A(r)$ with respect to r, and then set this equal to zero?

      iv. Did you solve for r in the equation: $4\pi r - \dfrac{2000}{r^2} = 0$? If not, see below.

   c. Many students have trouble solving for r in this equation. One student's solution is shown at right ➔

      i. Briefly describe what the student did in each step.

      ii. Is this solution valid?

      iii. Does this solution match the graph at the top of the page? Explain how you can tell.

$$4\pi r - \frac{2000}{r^2} = 0$$

$$4\pi r = \frac{2000}{r^2}$$

$$r^2 \cdot 4\pi r = \frac{2000}{\cancel{r^2}} \cdot \cancel{r^2}$$

$$r^2 \cdot 4\pi r = 2000$$

$$\pi r^3 = 500$$

$$r^3 = \frac{500}{\pi}$$

$$r = \left(\frac{500}{\pi}\right)^{\frac{1}{3}}$$

16. For each of the following, draw a picture, assign variables as you see fit, and generate an equation that can be maximized (or minimized) to answer the question. For now, just describe in words how you could find the solution using this equation. **For homework, provide a full solution.**

    a. A lifeguard is standing at the corner of a 50 m x 40 m rectangular swimming pool. A child at the exact center of the pool needs help. Assume the lifeguard always moves at a constant velocity on land (4 m/s), and a different constant velocity in the pool (2 m/s). (That is, neglect changes in velocity when he starts to run or swim, or when he dives into the pool.) If he wants to reach the child in the shortest time, how far along which edge of the pool should he run before he enters the pool and begins to swim directly toward the child?

    b. Sylvia wants an open top, rectangular fish tank with a base twice as long as it is wide. She doesn't care how tall it is, but it must have a volume of 30 cubic feet. Material for the base costs \$30 per square foot, while material for the sides costs \$20 per square foot. Find the cost of the cheapest tank with these specifications.

    c. Find the point (x, y) on the parabola $y = 0.5x^2$ that is closest to the point $(-2,4)$.

17. It turns out that, for a given volume, the cylinder with the minimum surface area has the property that $h = 2r$ (that is, the height is equal to the diameter). Show that this is the case for the 1 liter cylinder of minimum surface area in Question 14.

Notes

Integration 1: Area and Distance

Model 1: Luca's Bicycle Trip

Luca rides his bicycle due north along a road at a constant velocity of 20 kilometers per hour (km/h) for five hours, as modeled by $y = h(t)$.

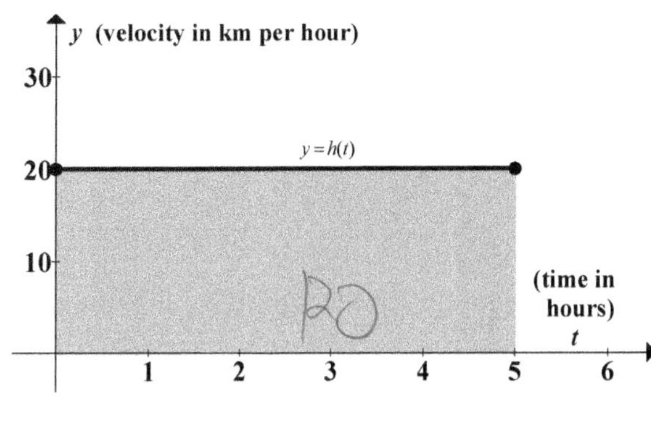

Construct Your Understanding

1. How far north did Luca travel during the 5 hour trip shown in Model 1? Include units in your answer.

2. For the graph in Model 1, calculate the area of the shaded rectangle. Show your work.

3. Use the units on the t and y axis to explain how your answers to Questions 1 and 2 are related.

4. The function $g(t)$, below right, models Allie's velocity during a bicycle trip.
   Note: Her real velocity function is continuous, but since (at $t = 2$) she accelerated from 15 to 20 km/h within a few seconds, we can use the discontinuous function below as a simplified representation of her trip. Look for a similar simplification on the next page.

   a. Shade the area between the graph of $g(t)$ and the horizontal axis.

   b. Calculate the area of this shaded region, and use units to explain what this value represents. Show your work.

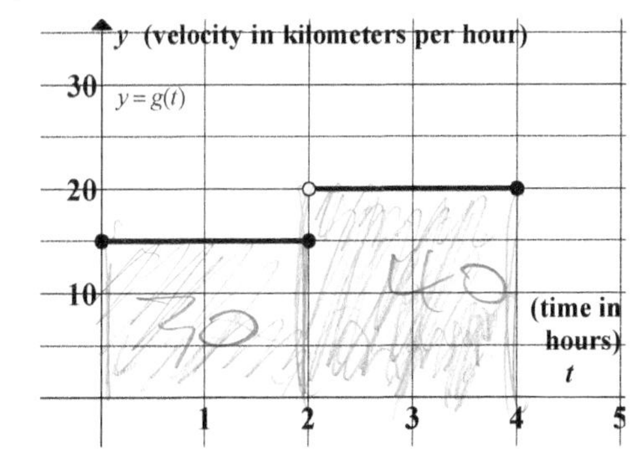

   c. (Check your work) Allie is 70 km from her starting point at the end of her four hour trip. Does this value match your answer to part b? If not, go back and determine how this value can be obtained by the analysis in part b.

5. Based on the examples we have seen so far, what information about a moving object is conveyed by the area between a velocity versus time curve and the horizontal axis?

6. Nico travels due north at 12 km/h for 2 hours, then turns around and travels due south at 20 km/h for 1 hour. Let velocity traveling north be positive and velocity traveling south be negative. Assume the changes in direction happen within a matter of seconds. Which function best approximates Nico's velocity during this three hour trip?

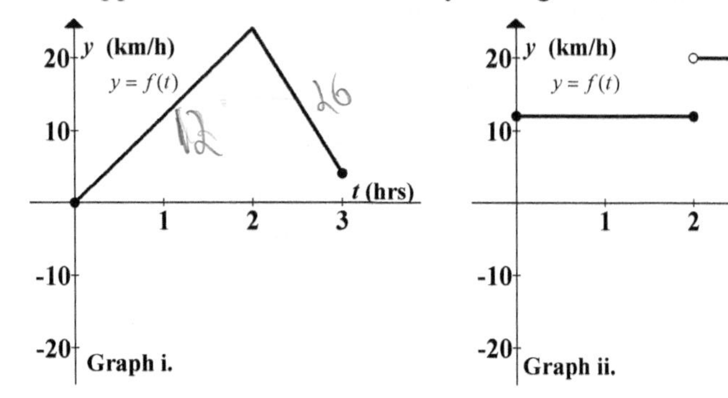

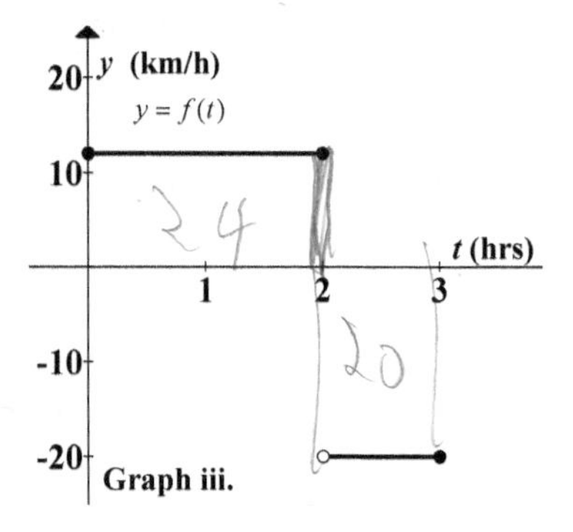

Briefly explain your reasoning.

7. For the graph you chose in the previous question…

   a. Shade the area between $f(t)$ and the horizontal axis for $t = 0$ to $t = 2$, and calculate the area of this shaded region.

      What does this value represent for this part of Nico's trip (include units)?

   b. Shade the area between $f(t)$ and the horizontal axis for $t = 2$ to $t = 3$, and calculate the area of this shaded region.

      What does this value represent for this part of Nico's trip (include units)?

   c. What do each of the following tell you about Nico's trip:

      i. (the area you shaded in part a) + (the area you shaded in part b)

      ii. (the area you shaded in part a) − (the area you shaded in part b)

8. (Check your work) Nico pedaled a total of 44 km, and ended up 4 km north of his starting point. Match each of these values with the appropriate part of your answer to Question 7c. If these values do not match either part, go back and check your work.

9. The function $h(t)$, at right, models velocity during a bicycle trip in which Andrei traveled north then south then north again along a road. The area of each shaded region is given on the graph (in parenthesis).

   At $t = 3$ hours, Andrei is __________ km north of his starting point. Explain your calculation.

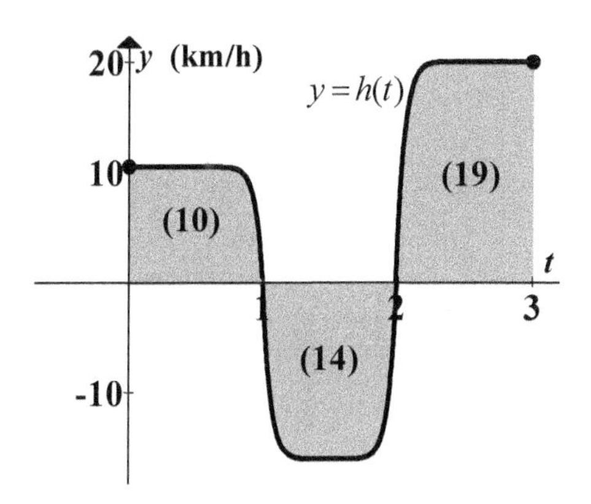

10. The function $v(t) = 20\cos 2t$ models the velocity of a particle moving back and forth in an electric field. The area of each shaded region is given in parenthesis (units = nanometers).

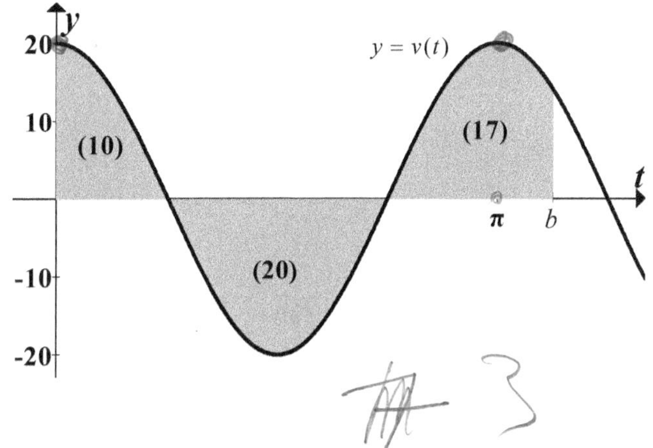

   a. At $t = b$, how far is the particle from where it was at $t = 0$?

   b. (Check your work) Is your answer to part a consistent with the fact that the particle has the same position at $t = 0$ and at $t = \pi$? Explain.

Model 2: Milo's Bicycle Trip

Milo's velocity is given by the function $f(t)$, shown as a smooth curve on the graph below.

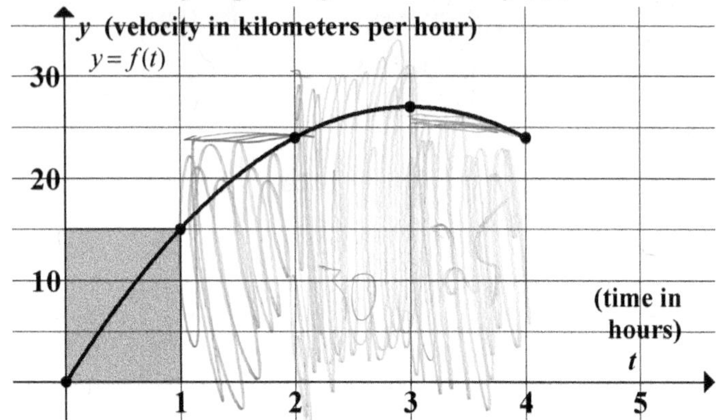

Instantaneous velocity readings are shown on the table below and as points on the graph.

Time (t)	Velocity (y)
0	0
1	15
2	24
3	27
4	24

Construct Your Understanding Questions

11. What is $f(0) =$ ____ $f(1) =$ ____ ?

    What do these values tell you about Milo's trip?

12. What is the area of the shaded rectangle in Model 2? If we take this as an estimate of the distance Milo traveled during the first hour, is this an **over**- or **under**-estimate? Explain.

13. The estimate described in the previous question assumes that Milo was traveling at a constant velocity of ________ km/h for the *whole* first hour of his trip.

14. (Check your work) Making this same type of estimate, we would assume Milo's velocity was 24 km/h for the *whole* second hour. Is this consistent with your answer to the previous question?

    a. Add a shaded rectangle to Model 2 representing Milo's change in position if he were going 24 km/h from $t = 1$ to $t = 2$.

    b. What is the area of this rectangle?

15. Making this same type of estimate, we would assume Milo's velocity was ________ km/h for the *whole* third hour.

    a. Add a shaded rectangle to Model 2 representing Milo's change in position if he were going this velocity from $t = 2$ to $t = 3$.

    b. Calculate the area of this rectangle.

16. Making this same type of estimate, we would assume Milo's velocity was ________ km/h for the *whole* fourth hour.

    a. Add a shaded rectangle to Model 2 representing Milo's change in position if he were going this velocity from $t = 3$ to $t = 4$.

    b. Calculate the area of this rectangle.

17. Calculate the sum of the areas of all four shaded rectangles in Model 2. State what this sum tells you about Milo's trip and include units in your answer.

18. Another way to estimate Milo's position at $t = 4$ is to assume he went…

    0 km/h for the 1st hour

    15 km/h for the 2nd hour

    24 km/h for the 3rd hour

    27 km/h for the 4th hour

    (as shown at right).

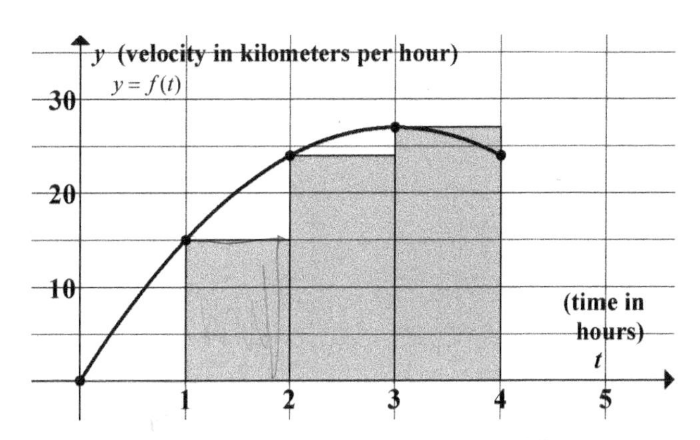

a. Estimate Milo's position at $t = 4$ using these rectangles, and explain your reasoning.

b. Describe how the estimate you made in this question differs from the estimate you made in Question 17. In particular, describe the type of velocity reading that was used in each.

Summary Box I1.2: Riemann Sums

The sums used in Questions 17 and 18 to estimate the area between the graph of a function f and the horizontal axis, for x values in an interval [a,b], are examples of **Riemann sums**.

If the height of each rectangle is the value of the function at the right-hand endpoint of the subinterval, then the sum is called a **right-hand Riemann sum**.

If the height of each rectangle is the value of the function at the left-hand endpoint of the subinterval, then the sum is called a **left-hand Riemann sum**.

One can similarly define **mid-point Riemann sums**. In fact, one can calculate a Riemann sum in which the heights of rectangles are equal to y values corresponding to any arbitrary x value in the given subinterval.

19. Based on the information in Summary Box I1.2, draw lines to match the beginning of each sentence with the correct middle and end to make a true statement about Milo's trip.

Beginning	Middle	End
The estimate in Question 17 uses the values…	$f(0)$, $f(1)$, $f(2)$, and $f(3)$ as the heights of the four rectangles…	representing change in position during each hour, and is an example of a **left hand Riemann Sum**.
The estimate in Question 18 uses the values…	$f(0.5)$, $f(1.5)$, $f(2.5)$, and $f(3.5)$ as the heights of the four rectangles…	representing change in position during each hour, and is an example of a **mid-point Riemann Sum**.
Another type of estimate (not seen in this activity) uses the values…	$f(1)$, $f(2)$, $f(3)$, and $f(4)$ as the heights of the four rectangles…	representing change in position during each hour, and is an example of a **right hand Riemann Sum**.

Notes

Integration 2: Riemann Sums

Model 1: Area and Distance (Review)

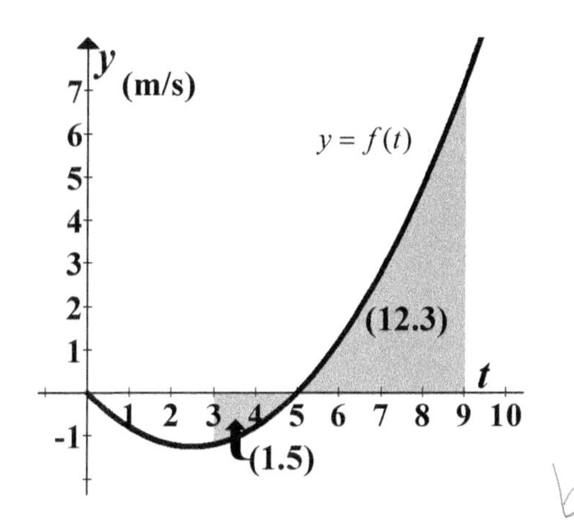

An object moves down (negative velocity) and then up (positive velocity) according to the velocity function $y = f(t)$, shown on the graph at right. Recall that change in position over a given interval can be calculated if you know the area between the graph of $y = f(t)$ and the t-axis on a given interval.

For example: The area between $y = f(t)$ and the t-axis from $t = 3$ to $t = 5$ is given as 1.5 m on the graph. Since this is an interval of *negative* velocity, this area tells us that the object moved 1.5 meters *down* during this interval.

Construct Your Understanding Questions

1. At what time does the object begin to move in an upward direction? At $t =$
2. How far does the object move in the upward direction between $t = 5$ and $t = 9$?
3. At $t = 9$ how far is the object from where it was at $t = 3$? Explain how to determine this from the areas given on the graph.

4. (Check your work) In the margin, draw a picture of the object's path during the interval from $t = 3$ to $t = 9$. Mark the point on your drawing at $t = 5$ and label the drawing with the distance the object travels in each direction (down and then up).
5. If we were not given the areas of the shaded regions in Model 1, we could use the rectangles of a **Riemann sum** to estimate them. For simplicity, let us focus on $t = 5$ to 9.

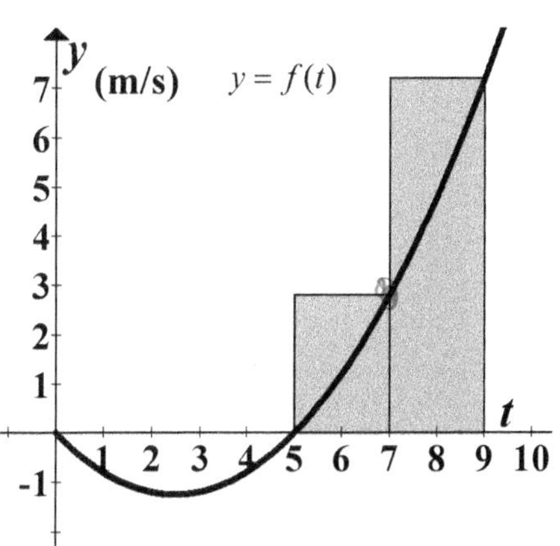

   a. What is the width (Δt) of each shaded rectangle on the graph at right?
   b. The function in Model 1 (and at right) is $f(t) = \frac{1}{5}t^2 - t$. Use this to calculate the height, and then the area of each rectangle.

   c. Sum the areas of these rectangles. This sum is an estimate of … (complete the sentence)

   d. Is this an **over** or **under**-estimate?
   e. (Review from CalcActivity I1) Is this a **left** or **right**-hand Riemann sum?

f. (Check your work) If a function is increasing on an interval then a right hand Riemann sum will give an overestimate and a left-hand Riemann sum will give an underestimate. Is this consistent with your answer to the previous question?

6. Use a right-hand Riemann sum with rectangles of width $\Delta t = 1$ to estimate the area between the graph of f and the horizontal axis over the interval from $t = 5$ to $t = 9$. Start by drawing the rectangles on the graph. Recall that $f(t) = \frac{1}{5}t^2 - t$. Show work.

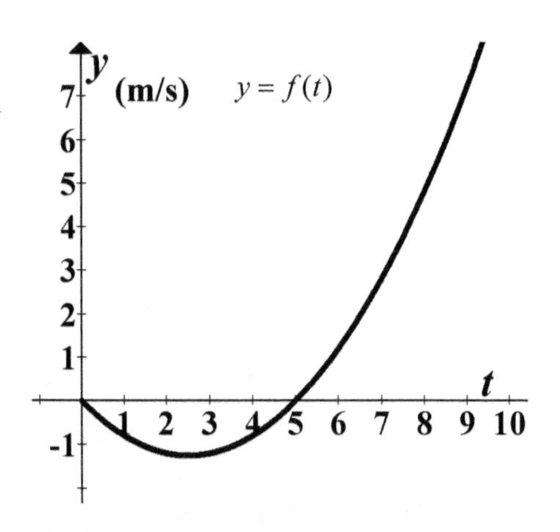

a. Compare this result to your previous estimate: Which is closer to the actual value given on the graph in Model 1?

b. What would be the effect on this estimate of further subdividing the rectangles you drew above into a larger number of rectangles? Explain.

7. The sum of the areas of the shaded rectangles on this graph constitutes a right-hand Riemann sum.

   a. What is $\Delta t =$

   b. If you were to sum the areas of these rectangles you would get a value of 14.1 m.

   (Check your work) Does this fit with your answer to part b of the previous question? Explain.

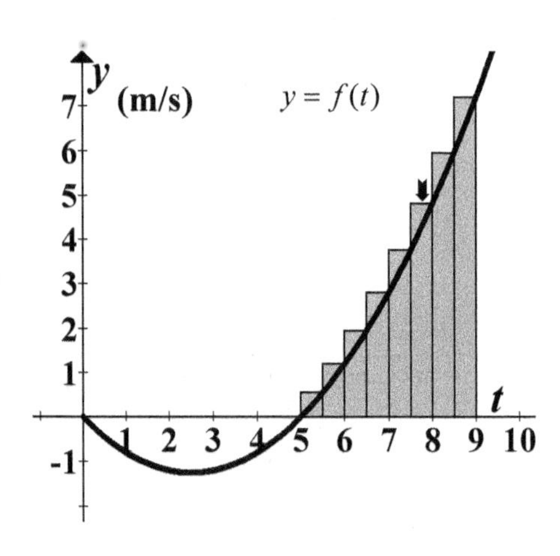

8. In order to write down a Riemann sum for a function $f(t)$ on an interval $[a,b]$ we have to partition this interval into smaller subintervals. To do this, we designate the left-hand endpoint of this interval (that is, the number a) with the symbol t_0. The numbers marking the boundary between each subsequent subinterval can then be written t_1, t_2, t_3, etc. This means that for the interval in the previous question, $t_0 = 5$, and $t_1 = 5.5$, $t_2 = 6$, $t_3 = 6.5$, etc..

   a. Which symbol gives the height of the first (farthest left) rectangle in a right-hand Riemann sum? t $\quad t_0$ $\quad t_1$ $\quad f(t)$ $\quad f(t_0)$ $\quad f(t_1)$ $\quad \Delta t$

b. Circle each expression that describes the area of the rectangle in the previous question that is marked with ⬇. Circle all that apply.

$f(7.5)\cdot 0.5$ $f(8)\cdot 0.5$ $f(t_8)\cdot \Delta t$ $f(t_7)\cdot \Delta t$ $f(t_6)\cdot \Delta t$ 2.4 m 4.8 m

c. (Check your work) There are three correct answers to part b of this question. Make sure that you circled all three.

d. Mark the expression that describes the sum of the areas of the eight shaded rectangles on the graph in Question 7.

$$f(t_0)\cdot\Delta t + f(t_1)\cdot\Delta t + f(t_2)\cdot\Delta t + f(t_3)\cdot\Delta t + f(t_4)\cdot\Delta t + f(t_5)\cdot\Delta t + f(t_6)\cdot\Delta t + f(t_7)\cdot\Delta t$$

$$f(t_1)\cdot\Delta t + f(t_2)\cdot\Delta t + f(t_3)\cdot\Delta t + f(t_4)\cdot\Delta t + f(t_5)\cdot\Delta t + f(t_6)\cdot\Delta t + f(t_7)\cdot\Delta t + f(t_8)\cdot\Delta t$$

e. (Check your work) The other expression shown in part d describes a left-hand Riemann sum. Explain how you can tell it is a left-hand Riemann sum.

f. Rewrite each sum in part d of this question using sigma notation. (See Summary Box I2.1, below.)

Summary Box I2.1: Sigma Notation

A sum can be expressed in shorthand using the Greek letter sigma. For example…

$$x_0 + x_1 + x_2 + x_3 + x_4 + x_5 = \sum_{i=0}^{5} x_i \quad \textbf{or} \quad f(t_1)+f(t_2)+f(t_3)+f(t_4)=\sum_{i=1}^{4} f(t_i)$$

$$ax_1 + ax_2 + \ldots + ax_{20} = \sum_{i=1}^{20} ax_i \quad \textbf{or} \quad f(t_0)\cdot\Delta t + f(t_1)\cdot\Delta t + f(t_2)\cdot\Delta t + \ldots + f(t_{n-1})\cdot\Delta t = \sum_{i=0}^{n-1} f(t_i)\cdot\Delta t$$

9. Let n be the number of rectangles represented in a Riemann sum between $x_0 = a$ and $x_n = b$ for the function $f(x)$, shown on the graph at right.

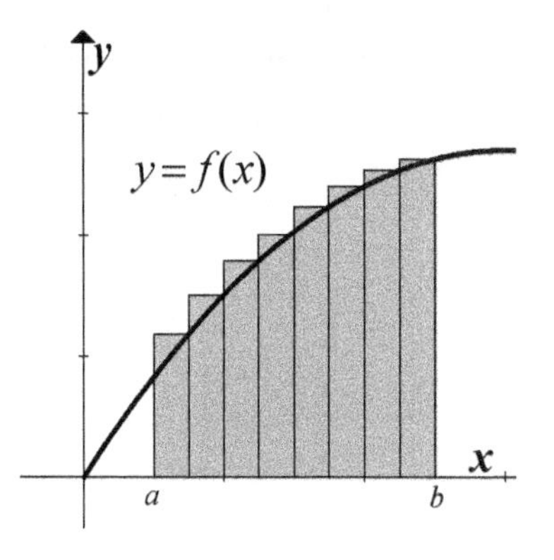

a. Which gives Δx, the width of each rectangle? (Assume the widths are all the same.)

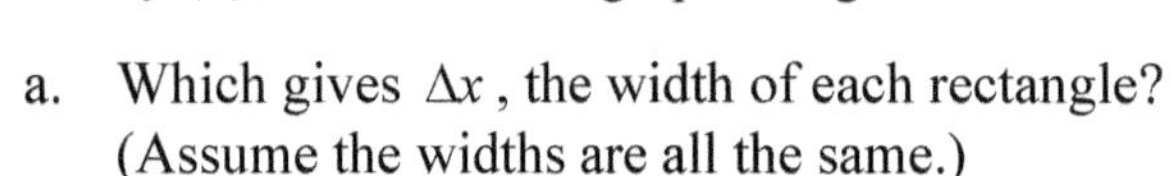

$b-a$ $\quad$ $a-b$

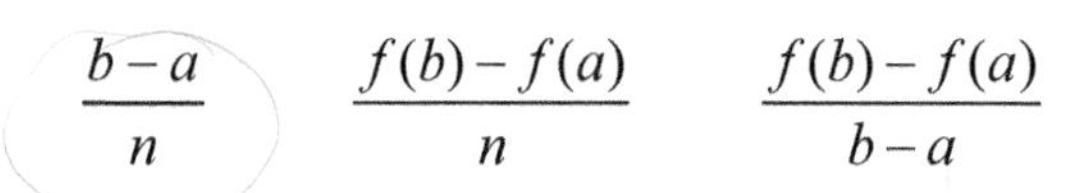

$\dfrac{b-a}{n}$ $\quad$ $\dfrac{f(b)-f(a)}{n}$ $\quad$ $\dfrac{f(b)-f(a)}{b-a}$

b. Given this expression for Δx, which expression gives the Riemann sum corresponding to the picture above, right?

$$\sum_{i=0}^{7} f(x_i)\cdot\Delta x \qquad \sum_{i=0}^{7} f(x_i+\frac{\Delta x}{2})\cdot\Delta x \qquad \sum_{i=1}^{8} f(x_i)\cdot\Delta x$$

Explain your reasoning.

c. Given this expression for Δx, which of the sums from part b (reprinted below) is the Riemann sum corresponding to the illustration on the graph at right?

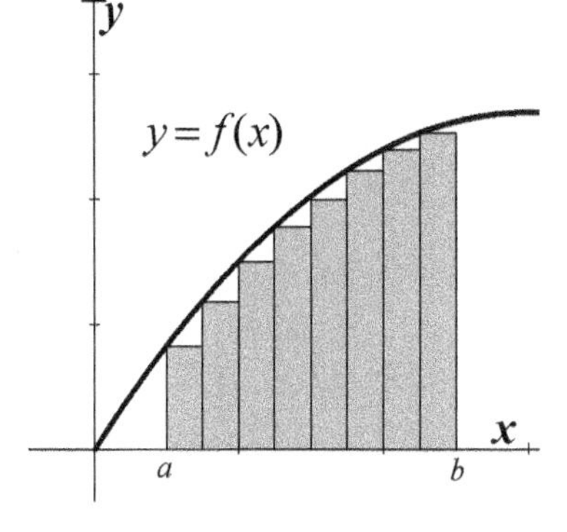

$$\sum_{i=0}^{7} f(x_i)\cdot\Delta x \qquad \sum_{i=0}^{7} f(x_i+\frac{\Delta x}{2})\cdot\Delta x \qquad \sum_{i=1}^{8} f(x_i)\cdot\Delta x$$

Explain your reasoning.

d. $\sum_{i=1}^{8} f(x_{i-1})\Delta x$ is equivalent to which sum shown in part c?

10. Label each Riemann sum in **parts c** and **d** of the previous question as a **left-hand sum, right-hand sum** or **neither**.

a. For the one that is neither, describe what value is used for the height of a given rectangle.

b. (Check your work) Are your answers to this question consistent with Summary Box I2.2?

Summary Box I2.2: Expression for Riemann Sums (with equal length subintervals)

If $f(x)$ is a function defined on $[a,b]$, and the interval $[a,b]$ is partitioned into subintervals $[x_{i-1}, x_i]$ $(1 \le i \le n)$ of equal† length given by $\Delta x = \dfrac{b-a}{n}$, where $x_0 = a$ and $x_n = b \ldots$

then a Riemann sum corresponding to this interval $[a,b]$ is given by…

$$\sum_{i=1}^{n} f(x_i^*)\cdot \Delta x \qquad \text{where } x_i^* \text{ is any number in the interval } [x_{i-1}, x_i].$$

If $x_i^* = x_{i-1}$, then the sum is a left-hand sum. If $x_i^* = x_i$ then the sum is a right-hand sum.

If $x_i^* = x_{i-1} + \dfrac{\Delta x}{2}$, then the sum is a ______________________ sum (fill in the blank).

†Note: The most general type of Riemann sum does not require the subintervals to be the same length.

11. As n, the number of rectangles in a Riemann sum, gets larger, what happens to the sum of the areas of the rectangles? (What does this sum approach?)

a. On the graph at right, shade an area equal to the value that all four Riemann Sums in Questions 9c and 9d approach as n gets very large.

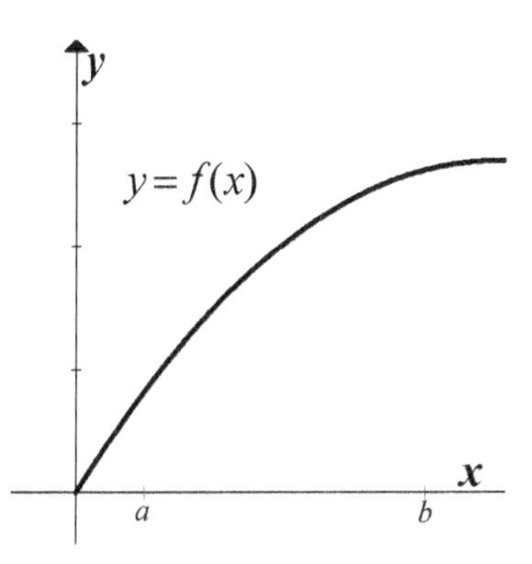

b. Express the area that you shaded at right as a limit of a Riemann sum.

12. (Check your work) Are your answers to the previous question consistent with Summary Box I2.3? If not, go back and revise your work.

Summary Box I2.3: Limit of a Riemann Sum

Given the partition of $[a, b]$ described in Summary Box I2.2, the *exact* area between the graph of a positive continuous function $f(x)$ defined on $[a,b]$ and the horizontal axis, for the interval from $x_0 = a$ to $x_n = b$, is equal to...

$$\lim_{n\to\infty} \sum_{i=1}^{n} f(x_i^*) \cdot \Delta x \quad \text{where } x_i^* \text{ is any number in the interval } [x_{i-1}, x_i], \text{ and } \Delta x = \frac{b-a}{n}.$$

Extend Your Understanding Questions (to do in or out of class)

13. For the Riemann sum pictured on the graph at right,

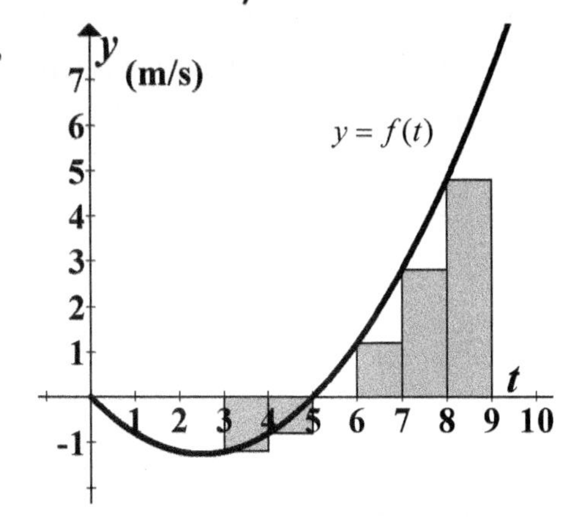

a. give values for...

i. $\Delta t =$

ii. $t_0 =$

iii. $n =$

b. Is this a **left** or **right**-hand Riemann sum? [circle one]

c. Write this sum using sigma notation.

d. Use the formula $f(t) = \frac{1}{5}t^2 - t$ to calculate the value of this sum. Show your work.

e. Given that this is the function from Model 1, describe what this value tells you about the moving object in Model 1?

f. Compare this value to the one you calculated in Question 3. Is this an **over** or **under** estimate?

14. On the graph at right, draw rectangles for a right-hand Riemann sum for the interval from $t_0 = 3$ to $t_n = 9$ and $n = 6$.

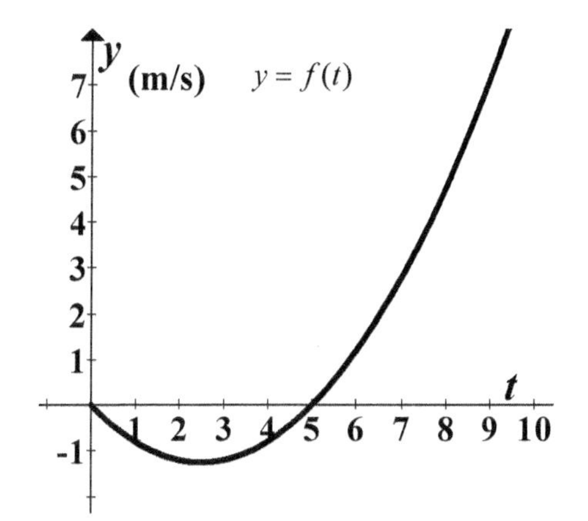

   a. Calculate the value of this sum.

   b. Compare this value to the one you calculated in Question 3. Is this an **over** or **under** estimate?

   c. Is f **increasing**, **decreasing**, or a **mixture** [circle one] over the interval from $t = 3$ to $t = 9$?

15. (Check your work) Recall that if a function is increasing over an interval then a right hand Riemann sum will give an overestimate and a left-hand Riemann sum will give an underestimate. Explain why this is the case using the examples graphed in the previous two questions.

16. Find the left, right and midpoint Riemann sums for…

   a. $f(x) = 5 - x^2$ on the interval $[0,6]$ for…

      i. $n = 2$

      ii. $n = 3$

      iii. $n = 6$.

   b. $f(x) = \sqrt{1+x}$ on the interval $[0,8]$ with…

      i. $\Delta x = 8$

      ii. $\Delta x = 2$

      iii. $\Delta x = 1$

Notes

Integration 3: The Definite Integral

Model 1: The Definite Integral as an Area

As we saw in the last activity, the area (L) of the shaded region of the graph of the positive continuous function below is equal to $\lim_{n\to\infty}\sum_{i=1}^{n} f(x_i^*)\cdot\Delta x = L$ where $\Delta x = \frac{b-a}{n}$, and x_i^* is any number in $[x_{i-1}, x_i]$, the interval that defines the base of the ith rectangle.

This area, L is equal to a quantity called the **definite integral** of $f(x)$ from a to b, and is represented...

$$\boxed{\int_a^b f(x)\,dx = L}$$

The expression in the box is read...

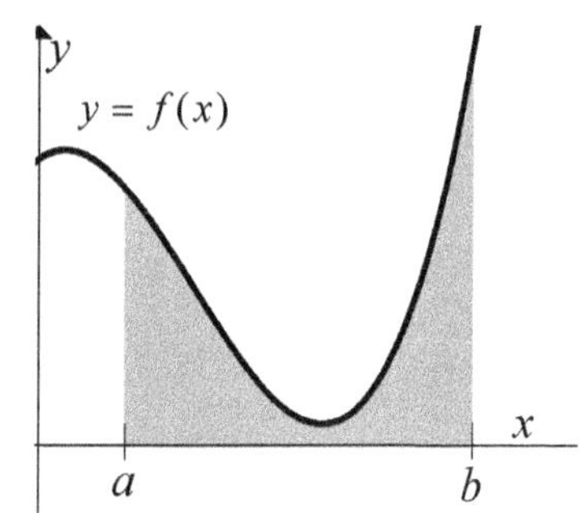

"The definite integral of f of x, d-x from x = a to x = b is equal to L."

The a and b next to $\int$ (the **integral sign**), are called the lower and upper **limits of integration**.

The integral sign is intended as a script S (for Sum) to remind us of the connection between the definite integral and Riemann sums. Because this limit L exists, the function is called **integrable**.†
$f(x)$ is called the **integrand**, and dx specifies that the independent variable is x.

† For functions that are not necessarily continuous, the definition of integrability is more subtle and requires the use of partitions of $[a,b]$ such that the subintervals do not necessarily have equal length.

Construct Your Understanding Questions (to do in class)

1. For the Riemann sum in Model 1, Δx gives the **height** or **width** [circle one] of each rectangle, and $f(x_i^*)$ gives the **height** or **width** [circle one] of the ith rectangle.

2. What symbol in the box in Model 1 showing the **definite integral** appears to be...
   a. analogous to the symbol Δx found in a Riemann sum?
   b. analogous to the symbol $f(x_i^*)$ found in a Riemann sum?

3. Which is equal to $\int_a^b g(x)\,dx$? [circle one]

   Area J   Area K   Area L   Areas J + K

   Areas K + L   Areas J + L   Areas J + K + L

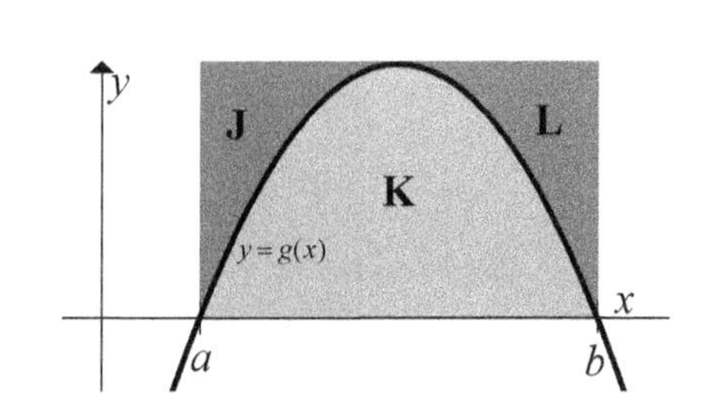

4. Which is equal to $\int_a^b f(x)\,dx$? [circle one]

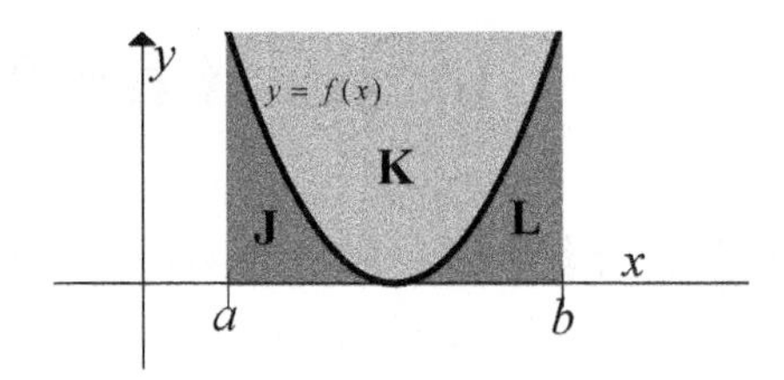

Area J Area K Area L Areas J + K

Areas K + L 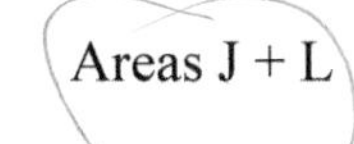 Areas J + L Areas J + K + L

5. For each graph, shade the area equal to the definite integral given.

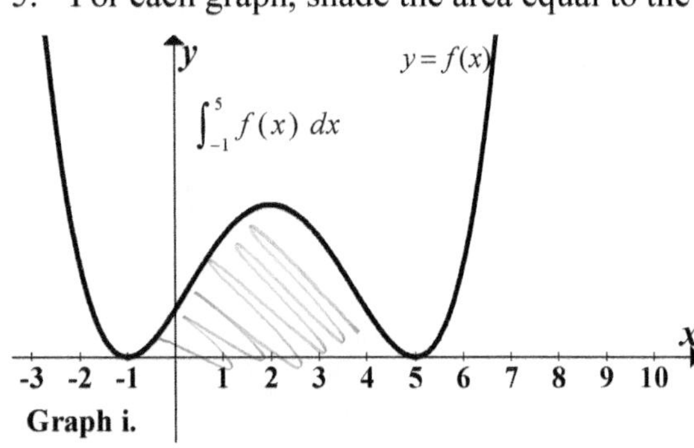

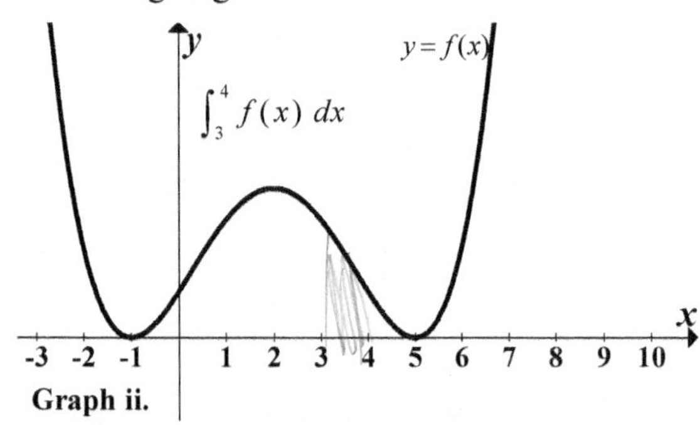

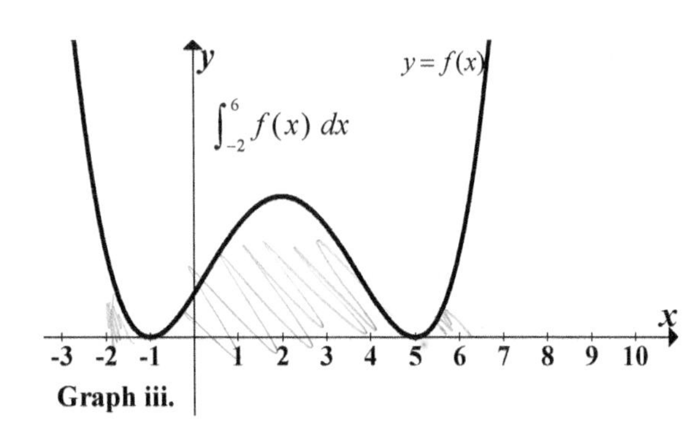

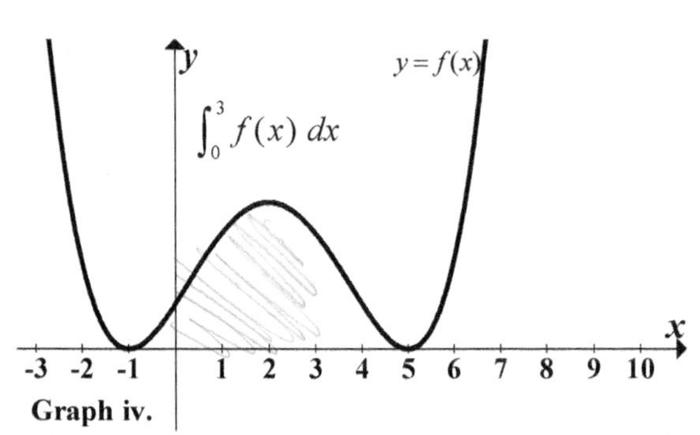

6. **A definite integral is a number**. For each definite integral in the previous question, this number is equal to the area that you shaded. Rank the four definite integrals in the previous question from largest to smallest.

7. For the function shown in Question 5…

   a. Give the <u>numerical</u> value of the definite integral…

      i. $\int_1^1 f(x)\,dx =$ ii. $\int_4^4 f(x)\,dx =$

   b. Write the definite integral from Question 5 that is equal to…
   (It may help to sketch $y = f(x)$ and shade the area equal to each integral.)

      i. $\int_{-1}^{1} f(x)\,dx + \int_{1}^{5} f(x)\,dx =$

      ii. $\int_{2}^{3} f(x)\,dx + \int_{0}^{2} f(x)\,dx =$

8. (Check your work) Are your answers to the previous question consistent with Summary Box I3.1? If not, go back and correct your work.

> **Summary Box I3.1: Some Properties of Definite Integrals**
>
> If a, b, c are real numbers and f is a continuous function:
>
> $\int_a^a f(x)\,dx = 0$ and $\int_a^c f(x)\,dx + \int_c^b f(x)\,dx = \int_a^b f(x)\,dx$

9. For **Graph v**, let x be time in seconds, and $f(x)$ be the velocity (in m/s) of an object that is moving to the left and then to the right.

   Use the areas of the shaded regions to find …

   a. At $x = 5$, how far is the object from where it was at $x = 0$?

   b. At $x = 5$, is the object to the **left** or **right** [circle one] of where it was at $x = 0$?

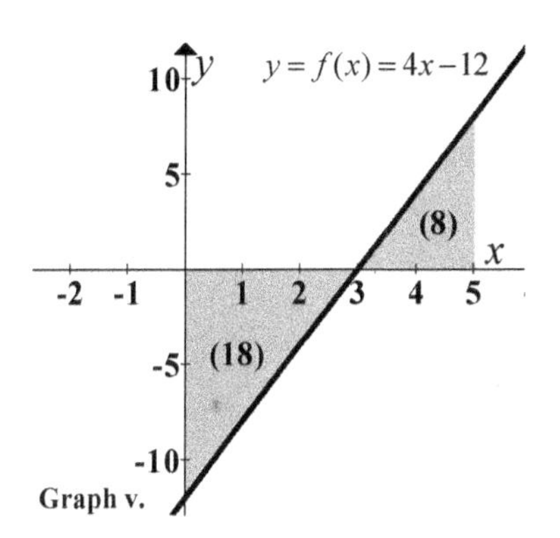

10. (Check your work) For $f(x)$ on **Graph v**, $\int_0^5 f(x)\,dx = -10$. Explain how both parts (a and b) of your answer to the previous question can be combined to give the value of this definite integral. If your answers do not match this value, go back and check your work.

11. For $f(x)$ on **Graph vi**, at right,

    find $\int_{-\pi}^{\frac{3\pi}{2}} f(x)\,dx =$

12. (Check your work) A student gives a value of 5 for the definite integral in the previous question. Explain the error this student is likely making.

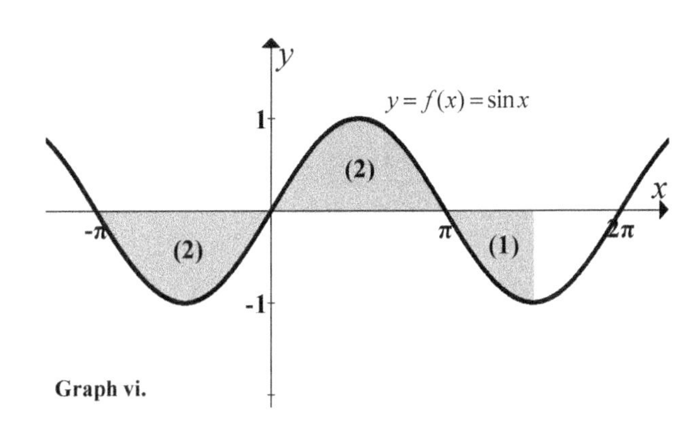

13. Use **Graphs v** and **vi** on this page to find the value of each definite integral.

    a. $\int_0^3 (4x - 12)\,dx =$ b. $\int_3^5 (4x - 12)\,dx =$ c. $\int_2^4 (4x - 12)\,dx =$

    d. $\int_0^{\frac{\pi}{2}} \sin x\,dx =$ e. $\int_{\frac{\pi}{2}}^{2\pi} \sin x\,dx =$ f. $\int_{-\frac{\pi}{2}}^{\pi} \sin x\,dx =$

14. For the constant function $f(x)=3$ shown on **Graph vii**, at right…

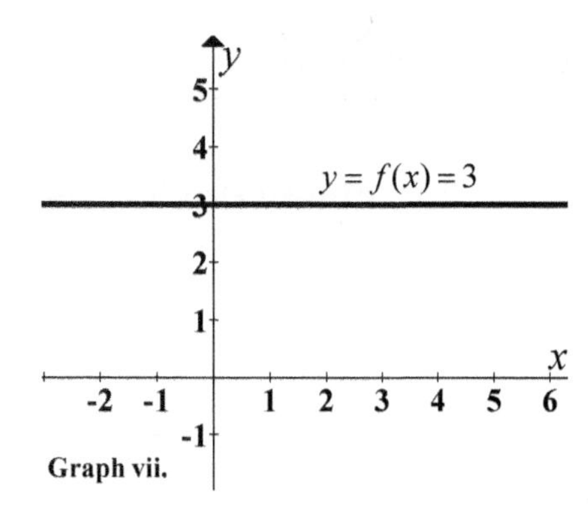

Graph vii.

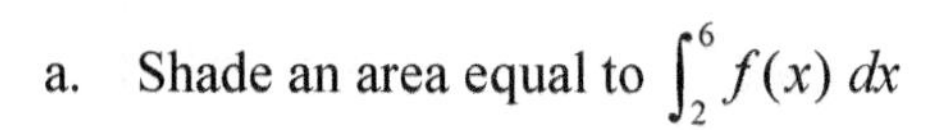
a. Shade an area equal to $\int_2^6 f(x)\,dx$

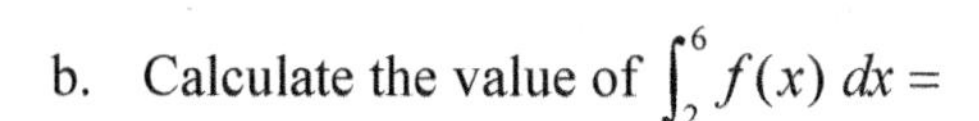
b. Calculate the value of $\int_2^6 f(x)\,dx =$

15. Consider the generalized constant function $f(x)=k$, where k is a constant (e.g., in the previous question $k=3$). In terms of k, a, and b, what is $\int_a^b k\,dx =$

16. (Check your work) For the function $f(x)=8$, the value of $\int_3^7 f(x)\,dx = 32$. Does your formula in the previous question give you this value? If not, check your formula.

17. Use the formula you derived in Question 15 to find the value of each definite integral.

a. For $f(x)=3$ (shown on **Graph vii**), what is $\int_6^2 f(x)\,dx =$

b. For $f(x)=8$ what is $\int_7^3 f(x)\,dx =$

18. (Check your work) For $f(x)=3$,
compare $\int_6^2 f(x)\,dx$ (in Question 17a),
with $\int_2^6 f(x)\,dx$ (in Question 14),
→ and describe how…

a. the limits of integration differ.

b. the values differ.

19. Based on what you have discovered so far, what can you say about the value of $\int_b^a k\,dx$ compared to the value of $\int_a^b k\,dx$? (k = a constant)

20. Recall that the definite integral, $\int_a^b f(x)\,dx = \lim_{n\to\infty}\sum_{i=1}^{n} f(x_i^*)\cdot\Delta x$, where $\Delta x = \dfrac{b-a}{n}$.

a. What is the sign of Δx when $b>a$?

b. What is the sign of Δx when $b<a$?

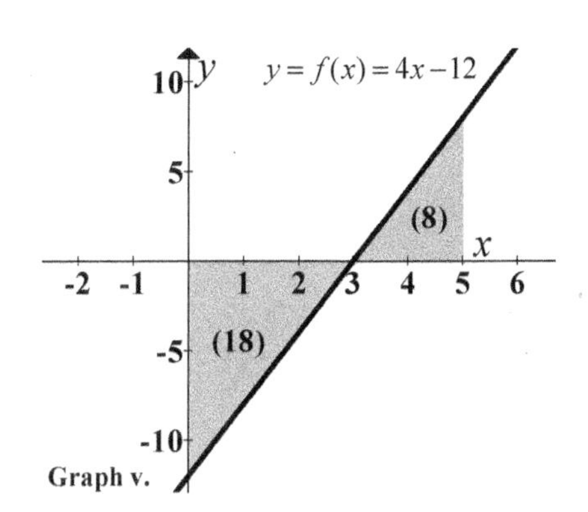

Graph v.

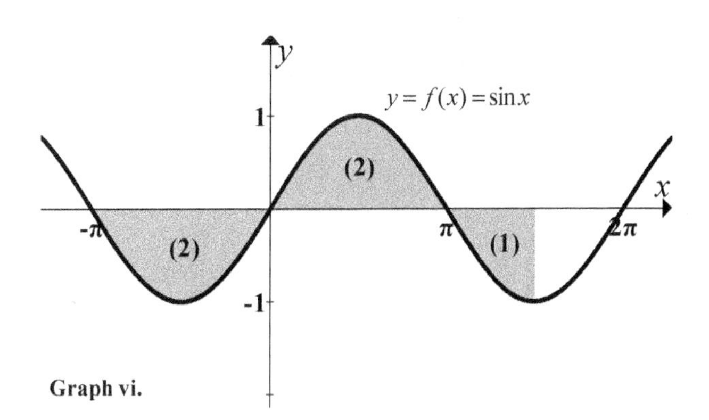

Graph vi.

21. Consider the following pairs of definite integrals along with selected values:

$\int_0^5 (4x-12)\,dx = -10$	$\int_0^\pi \sin x\,dx = 2$	$\int_1^4 x^2\,dx =$	$\int_a^b f(x)\,dx = L$
$\int_5^0 (4x-12)\,dx = 10$	$\int_\pi^0 \sin x\,dx = -2$	$\int_4^1 x^2\,dx = -21$	$\int_b^a f(x)\,dx =$

a. Determine the pattern, and guess the two missing values of the integrals above, based on this pattern.

b. Use this pattern and **Graphs v** and **vi** (reprinted above) to find…

i. $\int_3^0 (4x-12)\,dx =$ ii. $\int_\pi^{-\frac{\pi}{2}} \sin x\,dx =$

c. Explain how your answer to Question 20 is related to this pattern.

22. For $f(x)=3$, shown on **Graph vii** in Question 14, what is…

a. $\int_1^5 f(x)\,dx =$ b. $\int_1^5 7\cdot f(x)\,dx =$

23. (Check your work) Are your answers to Questions 15, 19, 21 and 22 consistent with Summary Box I3.2? If not, go back and check your work.

Summary Box I3.2: More Properties of Definite Integrals

If a and b are real numbers, k is a constant, and f is a continuous function:

$\int_a^b k\,dx = k(b-a)$ $\int_a^b f(x)\,dx = -\int_b^a f(x)\,dx$ $\int_a^b k\cdot f(x)\,dx = k\cdot\int_a^b f(x)\,dx$

24. Use **Graphs viii** and **ix** to find…

a. $\int_0^{2\pi} g(x)\,dx =$

b. $\int_0^{2\pi} 2\cdot g(x)\,dx =$

c. $\int_0^{2\pi} h(x)\,dx =$

d. $\int_0^{2\pi} [g(x)+h(x)]\,dx =$

e. $\int_0^{2\pi} [g(x)-h(x)]\,dx =$

f. $\int_0^{2\pi} [g(x)-3\cdot h(x)]\,dx =$

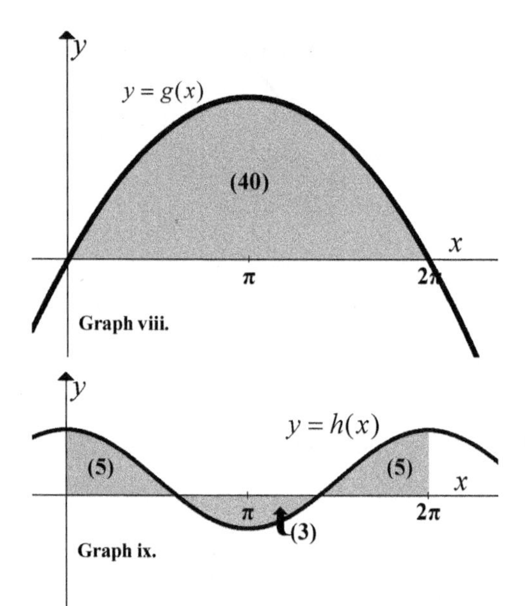

25. (Check your work) Are your answers to the previous question consistent with Summary Box I3.3? If not, go back and correct your work.

Summary Box I3.3: Addition, Subtraction or Multiplication by a Constant

$\int_a^b [f(x)+g(x)]\,dx = \int_a^b f(x)\,dx + \int_a^b g(x)\,dx$ (f and g integrable on $[a,b]$)

$\int_a^b [f(x)-g(x)]\,dx = \int_a^b f(x)\,dx - \int_a^b g(x)\,dx$ (f and g integrable on $[a,b]$)

Extend Your Understanding Questions (to do in or out of class)

26. One graph at right shows an even function, the other an odd function.

a. Label each.

b. Shade the area represented by  $\int_{-a}^{a} f(x)\,dx$ on each graph.

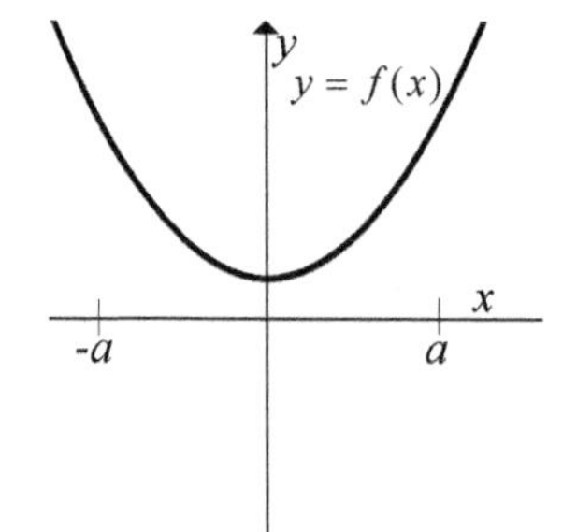

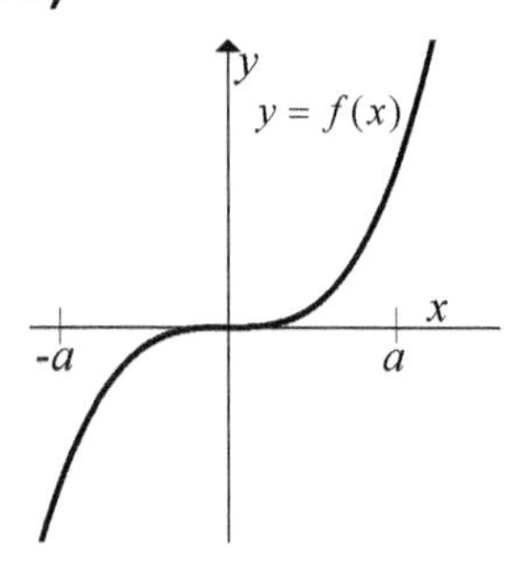

27. Assume $f(x)$ continuous on $[-a,a]$.

a. Which one of the following statements is true? (Mark the other false.)

i. **True** or **False**: For any even function, $\int_{-a}^{a} f(x)dx = 0$.

ii. **True** or **False**: For any odd function, $\int_{-a}^{a} f(x)dx = 0$.

b. Circle even or odd, and fill in limits of integration that make the statement true .

For an **even** or **odd** [circle one] function, $\int_{-a}^{a} f(x)dx = 2\cdot\int_{\square}^{\square} f(x)dx$

c. (Check your work) Are your answers to this question consistent with your labels and shading in the previous question? If not, check your work.

28. Find $\int_a^b f(x)\,dx$ when... Sketch a graph for each.

a. $a=-1$, $b=4$, $f(x)=x$

b. $a=-1$, $b=4$, $f(x)=\begin{cases} 2 & x\le 2 \\ x & x>2 \end{cases}$

29. Find $\int_a^b f(x)\,dx$ when... Sketch a possible graph for each.

a. $a=0$, $b=5$, and $f(x)=3$

b. $a=3$, $b=1$, and $f(x)=x$

c. $a=b$ and $f(x)$ is any function defined at $x=a$.

d. $a=1$, $b=5$, $\int_1^3 f(x)\,dx=12$, and $\int_3^5 f(x)\,dx=10$ f integrable on $[a,b]$

e. When $a=-1$, $b=5$, and $f(x)=2\cdot g(x)$ (Where $y=g(x)$ is shown on the graph at right.)

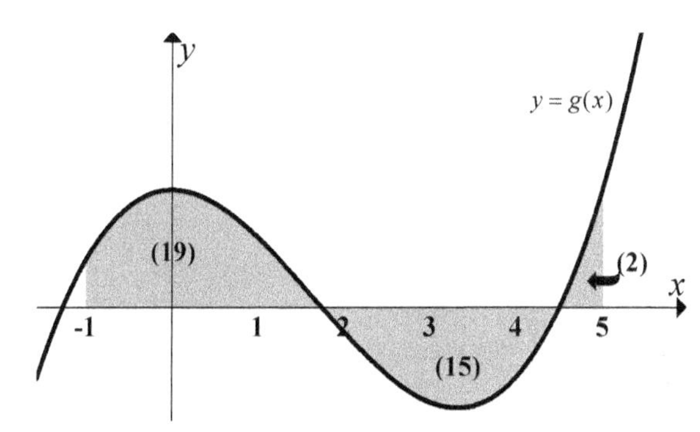

30. (Check your work) Each part of the previous question demonstrates a property listed in Summary Boxes I3.1-3. Match each part of the question to the correct property by writing each of the following labels next to the appropriate property in Summary Boxes I3.1-3.

Question 29a, Question 29b, Question 29c, Question 29d, and Question 29e

Notes

I4: Fundamental Theorem of Calculus (FTC)

Model 1: A High-Wire Act

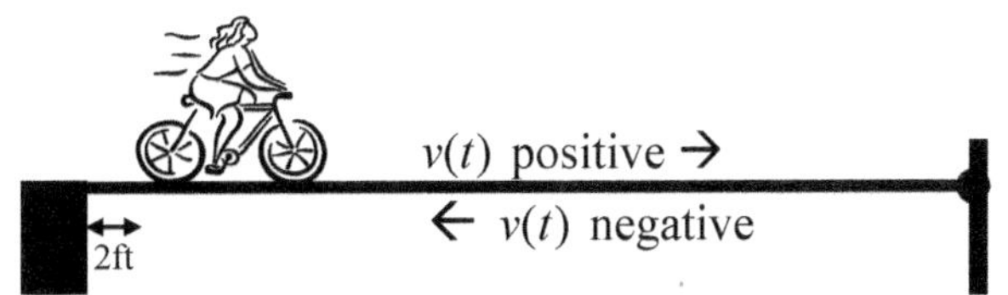

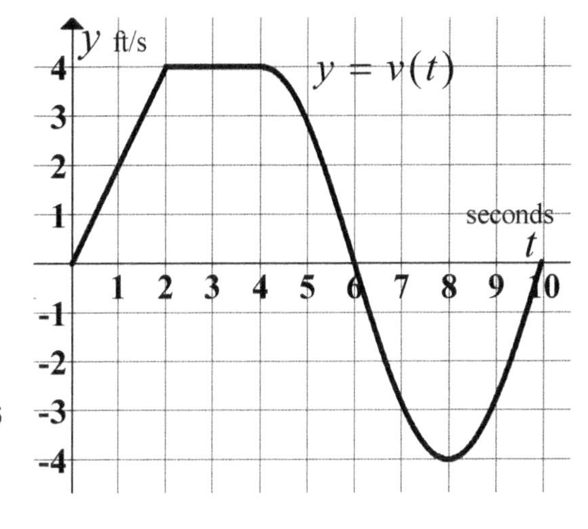

The graph at right plots the velocity $v(t)$ of a circus performer during a high-wire bicycle act. At $t = 0$ she is 2 feet from the starting platform.

Construct Your Understanding Questions

1. Based on the picture and the graph, over what time interval is the bicycle moving forward? …moving backward?

2. At $t = 0$ s, the bicycle's **position** = distance from the starting platform =_____ ft.

   At $t =$ ____s the bicycle is farthest from the starting platform and its velocity = ____ft/s.

3. (Check your work) Are your answers above consistent with the fact that at $t = 6$ s the bicycle is momentarily stationary? If not, check your understanding of Model 1.

4. Recall that when $v(t)$ is positive, the area between the graph of $y = v(t)$ and the t-axis from a to b is equal to distance traveled between $t = a$ and $t = b$.

   a. On the graph in Model 1 (above), shade the area equal to the distance traveled between $t = 0$ and $\boldsymbol{t = 2}$ **s**. Use the grid to estimate this value = _____feet.

   b. On the axes below-right, plot the bicycle's **position** at $t = 2$ s. Note that **position** is defined as the distance from the starting platform. (The point at $t = 0$ is given.)

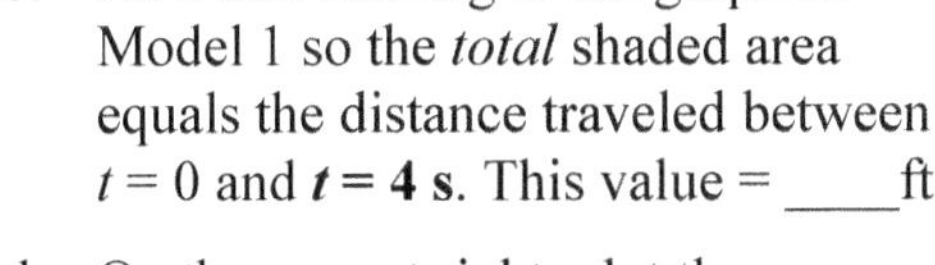

   c. Now add shading to the graph in Model 1 so the *total* shaded area equals the distance traveled between $t = 0$ and $\boldsymbol{t = 4}$ **s**. This value = ____ft

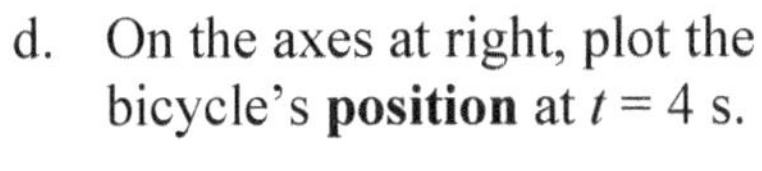

   d. On the axes at right, plot the bicycle's **position** at $t = 4$ s.

   e. Add shading to Model 1 so the *total* shaded area equals the distance traveled between $t = 0$ and $\boldsymbol{t = 6}$ **s**. Estimate this value = ____ft

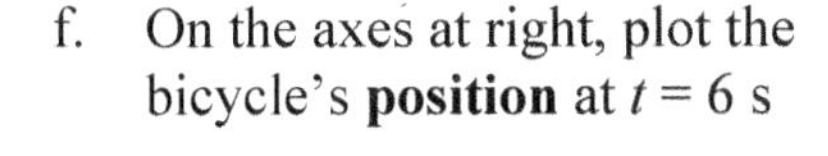

   f. On the axes at right, plot the bicycle's **position** at $t = 6$ s

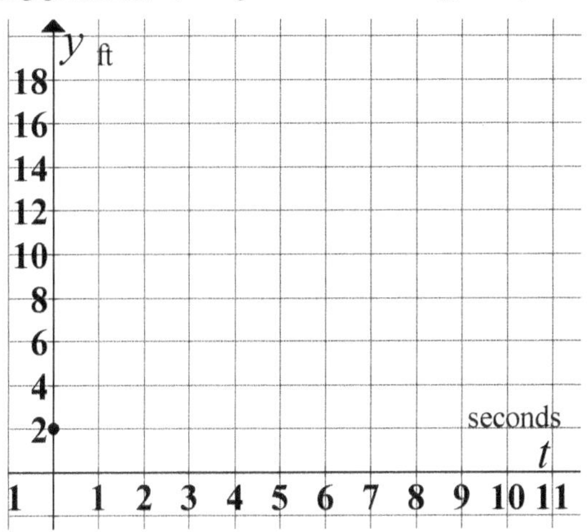

5. The definite integrals $\int_0^2 v(t)\,dt = 4$ and $\int_0^4 v(t)\,dt = 12$ are associated with parts a and c of the previous question. Give the definite integral (and value) that is associated with part e.

6. (Check your work) The position of the bicycle in Model 1, at $t = 6$ s is 19 ft.
   a. Is this consistent with your answer to Question 4f? If not, revise your work.
   b. The definite integral $\int_0^6 v(t)\,dt = 17$. Is this consistent with your answers to Questions 4e and 5? If not, go back and revise your work.
   c. Explain why $\int_0^6 v(t)\,dt$ is not equal to the position of the bicycle at $t = 6$ s.

7. In Model 1, shade the region between $y = v(t)$ and the t-axis from $t = 6$ to $t = 8$.
   a. Estimate the area of this newly shaded region (from $t = 6$ to $t = 8$) = _____ ft.
   b. During the period from $t = 6$ to $t = 8$ s, is the bicycle moving to the **left** or **right**?

8. At $t = 8$ seconds, the bicycle in Model 1 is 14 feet from the starting platform.
   a. Add a point to the position graph in Question 4 at $t = 8$ seconds.
   b. Based on this information, what is the value of…
      i. $\int_0^8 v(t)\,dt =$
      ii. $\int_6^8 v(t)\,dt =$
      The value of this integral (ii.) relates to both parts of Question 7. Explain.

9. For the bicycle in Model 1…
   a. Write down a definite integral that describes:
      i. The change in position between $t = 8$ and $t = 10$ s.
      ii. The change in position between $t = 0$ and $t = 10$ s.
   b. Use the grid in Model 1 to estimate the value of each these two definite integrals.
   c. Add a point to the graph in Question 4 at $t = 10$.

10. Sketch a curve through the points you plotted on the graph in Question 4. Label this new function $y = s(t)$, and describe what the function $s(t)$ tells you about the bicycle in Model 1.

11. (Check your work) Check that your curve matches the following values of $s(t)$.

$s(0) = 2$ $s(4) = 14$ $s(6) = 19$ $s(8) = 14$ $s(10) = 9$

12. For each definite integral listed below...
    a. Shade the graph so it represents the definite integral. The first one is done for you.
    b. Give each definite integral in terms of one or more of the values of $s(t)$ listed in the previous question.

Definite Integral	Shading Representing this Definite Integral	Definite Integral in Terms of $s(t)$
$\int_0^6 v(t)\,dt = 17$	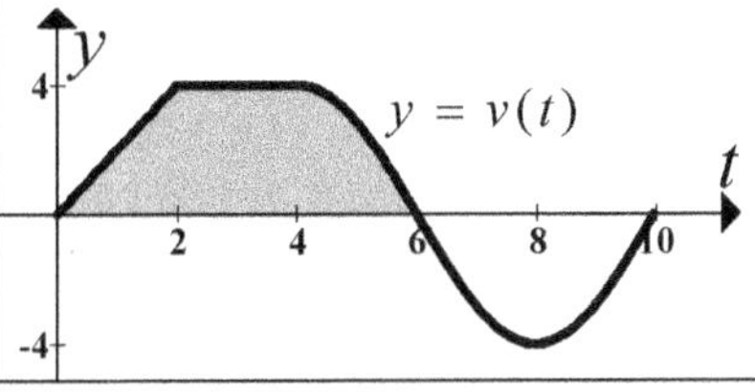	Hint: For this first one, it may help to review your answer to Question 6c.
$\int_0^8 v(t)\,dt = 12$	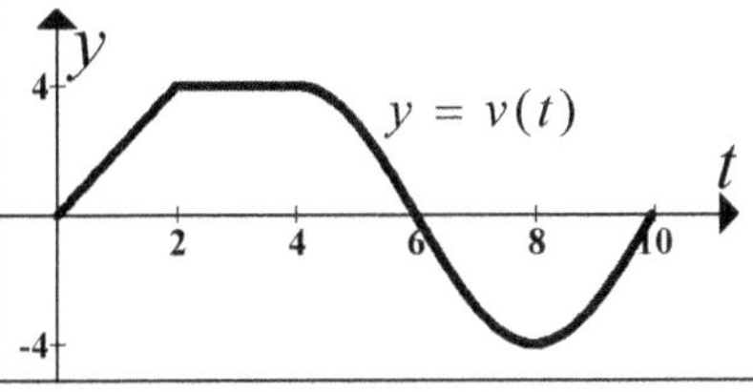	
$\int_4^6 v(t)\,dt = 5$	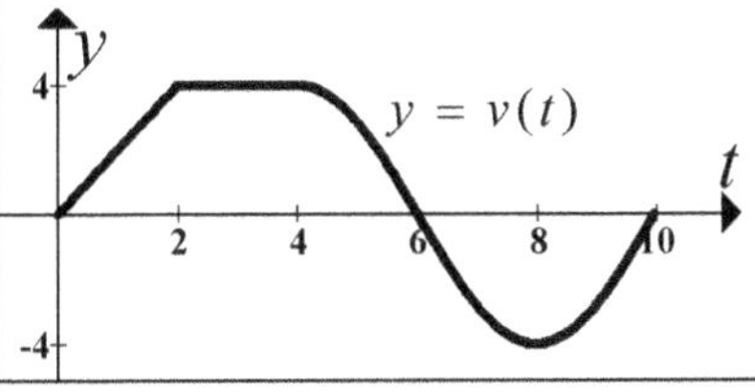	
$\int_8^{10} v(t)\,dt = -5$	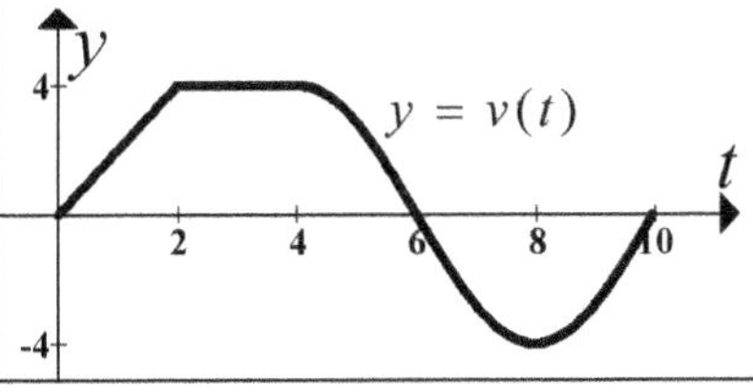	

13. For the function $s(t)$ which describes the position of the bicycle in Model 1, what is the value of the difference... $s(6)-s(4)=$

    a. Describe what this difference (and value) tells you about the high wire act in Model 1.

    b. Shade an area on the graph of $y=v(t)$, at right, that is equal to $s(6)-s(4)$, and explain your reasoning.

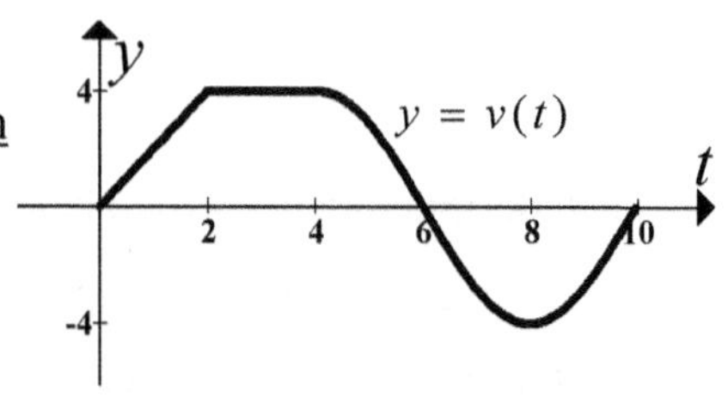

    c. Write down a definite integral that is equal to $s(6)-s(4)$, and explain your reasoning.

14. (Check your work) Identify the table entry in Question 12 that matches your answer to Question 13. If none match, discuss your answers with your group and write down a question for the instructor.

15. For v and s describing the motion of the bicycle in Model 1, and a in $[0,10]$...

    a. Give each definite integral in terms of a difference between two values of the function $s(t)$. The first one is done for you.

        i. $\int_4^6 v(t)\,dt = s(6)-s(4)$

        ii. $\int_3^6 v(t)\,dt =$

        iii. $\int_3^4 v(t)\,dt =$

        iv. $\int_a^b v(t)\,dt =$

b. Describe in words what the last definite integral in part a of this question (iv.) tells you about the motion of the bicycle in Model 1. Be prepared to share your group's answer as part of a whole-class discussion.

16. Based on what you have learned about derivatives in the first part of this course, which is a better description of the relationship between a velocity function and a position function, such as $v(t)$ and $s(t)$ for the bicycle in Model 1? $s = v'$ or $v = s'$ [Circle one.]

17. Neglecting air resistance, the position of a stone dropped from 100 m at $t = 0$ seconds (measured in meters from the surface), is given by the function: $s(t) = 100 - 5t^2$

(Check your work) Use your answer to the previous question to show that the velocity of the rock is given by the function: $v(t) = -10t$

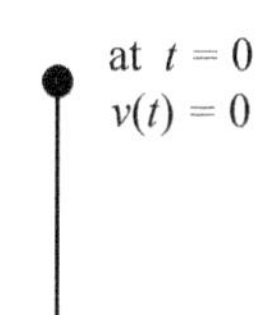

18. (Check your work) Are your answers to the last three questions consistent with Summary Box I4.1?

Summary Box I4.1: Fundamental Theorem of Calculus as Applied to Velocity

Let v be the velocity function of an object in motion, continuous on $[a,b]$, and let s be the object's position function. Then $s' = v$, and ...

$$\int_a^b v(t)\,dt = s(b) - s(a)$$

This says that the definite integral of the velocity function, over a given interval, gives the change in position during that interval.

Note: A general statement of the Fundamental Theorem of Calculus (often called the FTC) appears in the next activity.

19. Let us assume that the velocity (in meters/second) of an object is described by $v(t) = 2t$.

    a. Which function could be the position function for this object?

    i. $s(t) = 2$ ii. $s(t) = t^2 + 3$ iii. $s(t) = t + 2$

    Circle one and explain your reasoning:

    b. Construct a definite integral describing the change in position of this object during the interval from $t = 1$ to $t = 3$ seconds.

    c. Use the Fundamental Theorem of Calculus (FTC) to find a value for this definite integral. [Show your work.]

Extend Your Understanding Questions (to do in or out of class)

20. Given the function $v(t)$, use the FTC to find the value of the definite integral.

    a. Given $v(t)=t$, find $\int_2^5 v(t)\,dt =$

    Show your work.

    b. Given $v(t)=-7$, find $\int_0^3 v(t)\,dt =$

    Show your work.

21. (Check your work) Sketch a graph of the function…

    a. $v(t)=t$, and find the area between the function and the t-axis from $t=2$ to 5.
    Is this value consistent with your answer to part a of the previous question?

    b. $v(t)=-7$, and use this to check your answer to part b of the previous question.

22. For v and s in Model 1 and a in $[0,6]$…

    True or **False**: The area of the shaded region shown on the graph at right is equal to $s(a)$.

    Circle one and explain your reasoning.

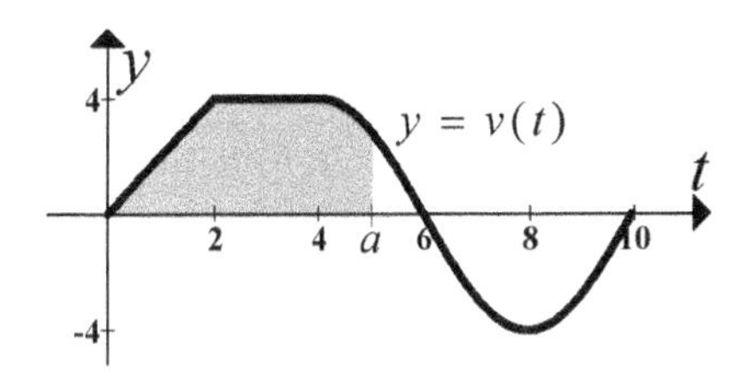

23. For v and s in Model 1 and a in $[0,10]$…

    a. Describe what the difference $s(a)-s(0)$ tells you about the bicycle.

    b. Write a definite integral equal to $s(a)-s(0)$.

24. (Check your work) Write a definite integral equal to the area of the shaded region in Question 22. Does this support your answer to Question 22? Explain.

Notes

I5: Antiderivatives and the Fundamental Theorem of Calculus

Model 1: A Graph of Velocity

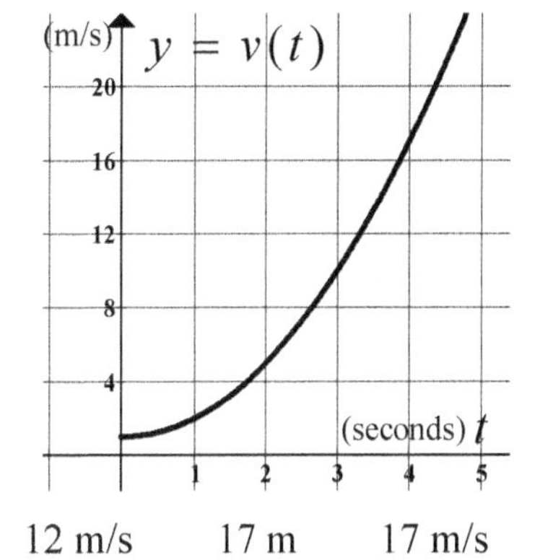

The graph of $y = v(t)$, at right, gives the velocity of a moving object in m/s as a function of time.

Construct Your Understanding Questions

1. On the graph, shade the area equal to $\int_0^3 v(t)\,dt$

   a. Use the grid to decide which of the following is closest to the value of this definite integral.

   3 m   3 m/s   8 m   8 m/s   12 m   12 m/s   17 m   17 m/s

   b. What does this value tell you about the movement of the object in Model 1?

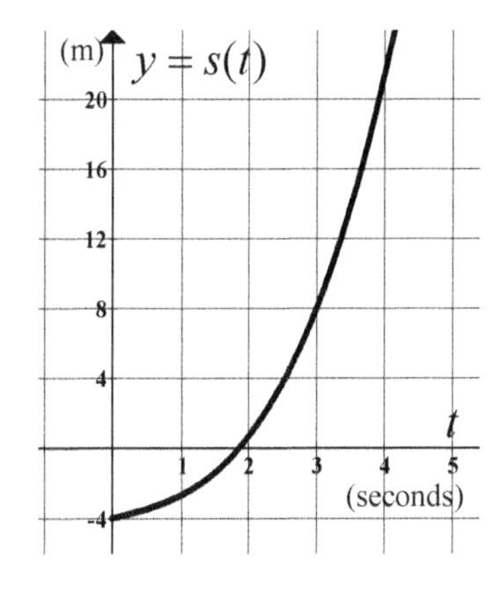

2. The graph of $y = s(t)$, below, right, gives the position of the object in Model 1.

   a. Use the graph of $y = s(t)$ to check your answer to Question 1a. Show your work and explain your reasoning.

   b. (Check your work) In answering part a did you find values for $s(3) =$ _____ and $s(0) =$ _____ and their difference? If not, read these values off the graph of $y = s(t)$, calculate their difference, and reformulate your answer to part a.

3. If v is a continuous function on $[a, b]$ that describes the velocity of a moving object, and s is the function that describes the position of that object, then, according to the FTC…

   a. $\int_a^b v(t)\,dt =$ $s(b) - s(a)$ or $s(a) - s(b)$ [circle the correct ending]

   b. $s' = v$ or $v' = s$ [circle one]

4. Suppose you know the formula for the velocity function of a moving object: $v(t) = t^2 + 1$.

   Describe how you can find the value of $\int_0^3 v(t)\,dt$ *without* a graph of $y = v(t)$ or $y = s(t)$. (If you wish, use your method to find the value of this definite integral.)

5. (Check your work) Does your answer to the previous question make use of the position function for the object, $s(t)$? If not, go back and revise your response.

6. (Review) Which function could be the position function of the object in Question 4?

   $s(t)=\frac{1}{3}t^3+t-4$ $s(t)=t^3+t$ $s(t)=2t$ [circle one]

7. (Check your work) For the moving object in Question 4, $\int_0^3 v(t)\,dt=12$. Show that your choice for $s(t)$ in the previous question gives a value of 12 for this definite integral.

8. Draw a line from each function $f(x)$ to the function $F(x)$ such that $F'(x)=f(x)$.

$f(x)=-10x$	$F(x)=10x+3$
$f(x)=10$	$F(x)=-5x^2$
$f(x)=3x^2$	$F(x)=x^3+17$
$f(x)=10x^{-2}$	$F(x)=2x^5-1$
$f(x)=10x^4$	$F(x)=-10x^{-1}$

9. Under each boxed function $f(x)$, supply a function $F(x)$ such that $F'(x)=f(x)$. The first one is done for you.

   a. $\boxed{f(x)=2x}$ $F(x)=x^2$ b. $\boxed{f(x)=x^2}$ $F(x)=$ c. $\boxed{f(x)=7x^6}$ $F(x)=$ d. $\boxed{f(x)=-2x^{-3}}$ $F(x)=$

10. (Check your work) Check your answers to the previous question by taking the derivative of each function $F(x)$, and making sure this is equal to the function $f(x)$ above it.

11. Fill in one blank with "**derivative**" and the other with the new term "**antiderivative**".

    If $F'(x)=f(x)$, then $f(x)$ is the ______________________ of $F(x)$, and $F(x)$ is a ______________________ of $f(x)$.

12. Fill in one blank with "**velocity**" and the other with "**position**".

    For a moving object, the ______________________ function is an antiderivative of the ______________________ function.

13. (Check your work) Are your answers to the previous two questions consistent with the information in Summary Box I5.1?

Summary Box I5.1: Fundamental Theorem of Calculus (General)

Let f be any function, continuous on $[a,b]$, and F be an antiderivative of f (that is, $F' = f$).

Then... $$\int_a^b f(x)\,dx = F(b) - F(a)$$

Notes: The Fundamental Theorem of Calculus is called FTC for short. There is another part to the FTC which you will encounter later.

14. Oil is leaking out of a ruptured tanker at a rate of $r(t) = 100e^{-\frac{t}{20}}$ liters per minute.

  a. What is the rate of oil flow at $t = 0$? .... at $t = 20$?

  b. Write down an expression that involves a definite integral (you need not solve this expression yet) that is equal to the amount of oil that leaked into the ocean between $t = 0$ and $t = 20$.

  c. Which one of the following functions is useful for solving the integral in part b?

  $R(t) = 100\cdot(-0.05)e^{-\frac{t}{20}}$ $R(t) = 100\cdot(-20)e^{-\frac{t}{20}}$ $R(t) = -5e^{-20t}$

  [Circle one and explain your reasoning.]

  d. (Check your work) For the function you chose in part c, show that $R'(t) = r(t)$.
  If this is not true, go back and reexamine your choices.

  e. Find the value of the definite integral in part b.
  Show your work.

15. Baby Bear's porridge cools at a rate given by the function $h(t)$ in degrees C/sec.

    a. Give the units of $\int_a^b h(t)\, dt$.

    b. Let $H(t)$ be an antiderivative of $h(t)$. Describe in words what information is conveyed by the difference, $H(b)-H(a)$.

16. The function $g(t)$ gives the rate at which guests enter an amusement park in people per hour.

    a. Give the units of $\int_a^b g(t)\, dt$

    b. Let $G(t)$ be an antiderivative of $g(t)$. Describe in words what information is conveyed by the difference, $G(b)-G(a)$.

17. Consider a rock, falling freely (no air resistance) after it was dropped from a 100 meter tower at $t=0$ seconds, so the distance from the ground is given by: $s(t)=-5t^2+100$.

    a. What function $v(t)$ describes the velocity of the rock? Explain your reasoning.

    b. Write the definite integral equal to the change in position of the rock during the interval $[2,4]$, and then find the value of this integral. (Show your work.)

Extend Your Understanding Questions (to do in or out of class)

18. The graph at right shows $y = v(t)$ and $y = s(t)$ for the rock in the previous question. Label each.

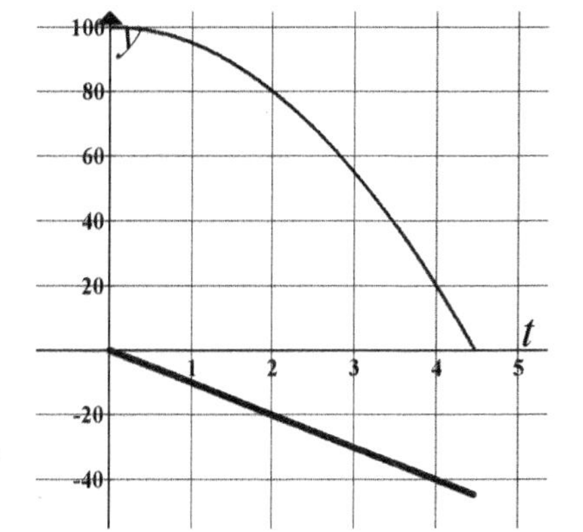

   a. Shade the area associated with the definite integral: $\int_2^4 v(t)\, dt$

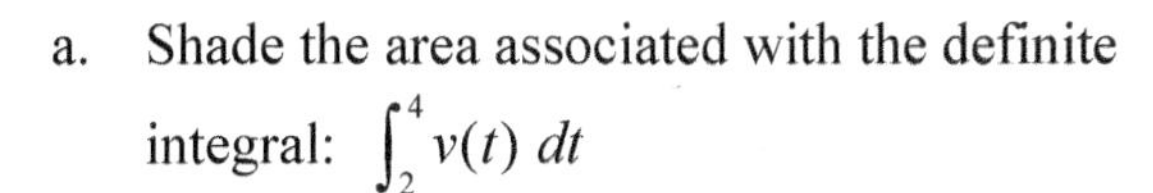

   b. Mark the points $(2, s(2))$ and $(4, s(4))$.

   c. There are two ways to check your answer to Question 17b: one uses the graph of $y = s(t)$, the other uses the graph of $y = v(t)$.

   Explain both types of checks, then confirm your answer to Question 17b with each.

19. Use the Fundamental Theorem of Calculus to evaluate (find the value of) $\int_a^b f(x)\, dx$. Show your work.

   a. $f(x) = 10$, $a = 6$, $b = 600$

   b. $f(x) = -10x$, $a = 4$, $b = 2$

   c. $f(x) = 3x^2$, $a = 2$, $b = 3$

   d. $f(x) = 12x^3$, $a = 2$, $b = 0$

Notes

Integration 6: Indefinite Integrals

Model 1: Graphs of Functions

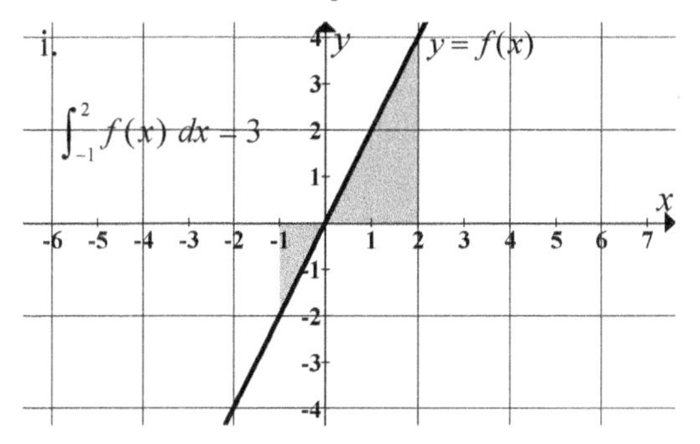

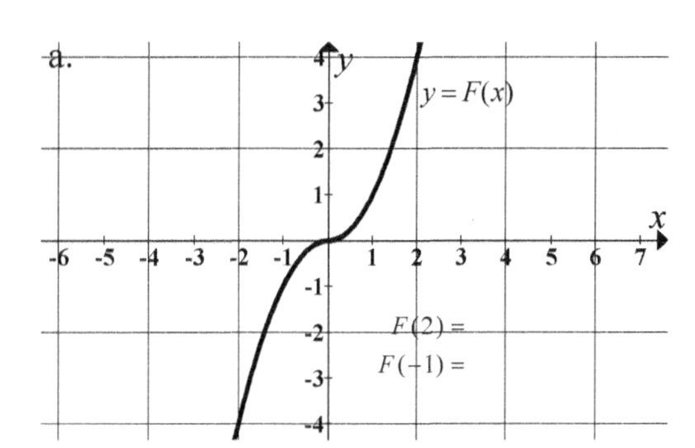

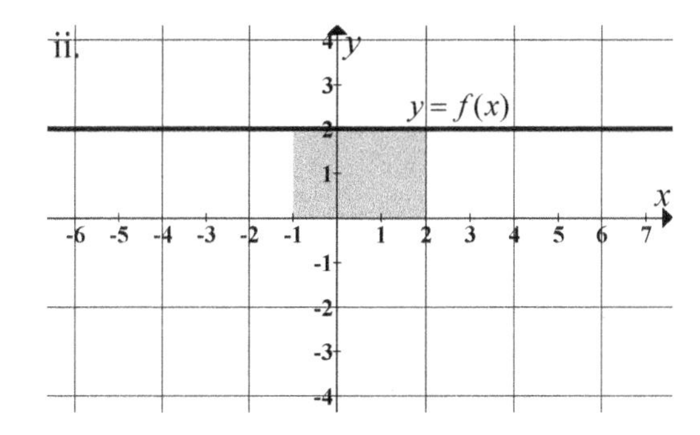

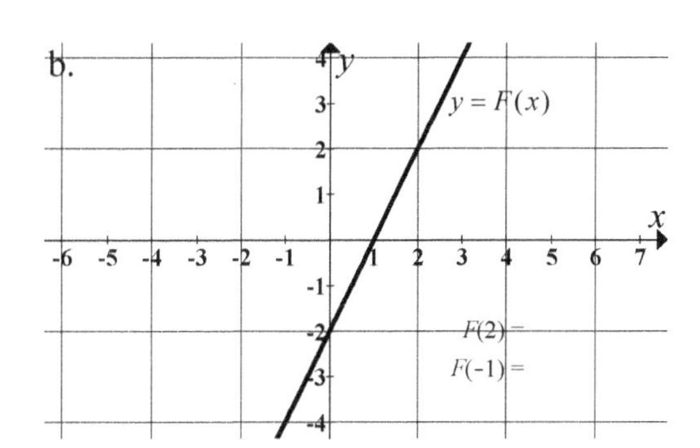

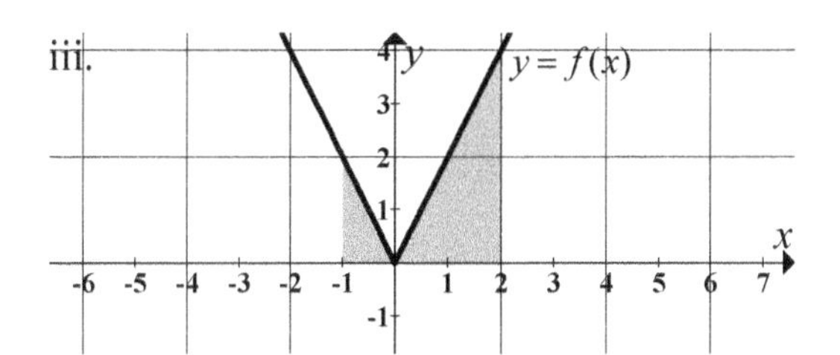

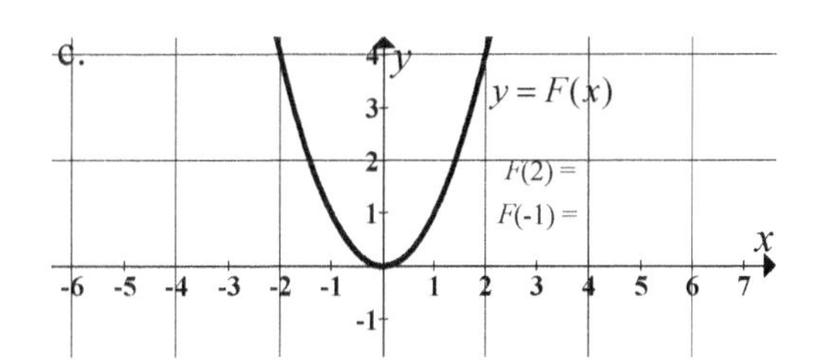

Construct Your Understanding Questions (to do in class)

1. On the graphs labeled i-iii (**on the left** in Model 1) write the definite integral associated with the shaded area shown, and estimate its value. The first one is done for you.
2. For Graphs a-c (**on the right** in Model 1) estimate the value of $F(2)$ and the value of $F(-1)$, and write these on the graph.
3. Recall that if F is an antiderivative of f, then f is the derivative of F (that is, $F' = f$). Draw a line from each function f (Graphs i-iii) on the left in Model 1 to the graph of F that is an antiderivative of f (choose from Graphs a-c, on the right of Model 1).
4. (Check your work) The Fundamental Theorem of Calculus (FTC) states that if f is continuous on $[a,b]$ and F is an antiderivative of f, then $\int_a^b f(x)\,dx = F(b) - F(a)$. Use the FTC to confirm your pairings in the previous question. Show your work.

5. Read $F(2)$ and $F(-1)$ from each graph, and record these on the graph.

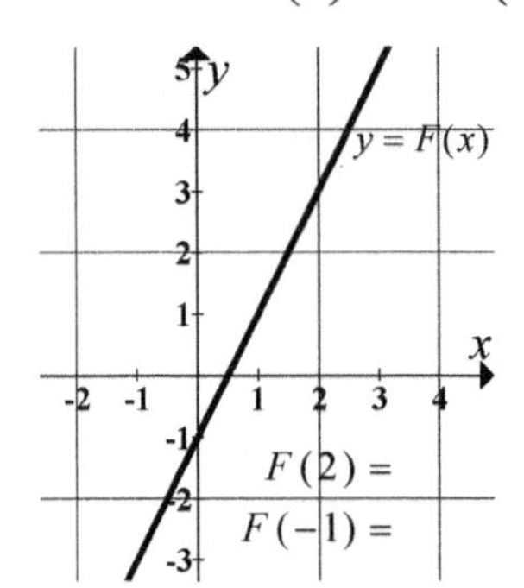

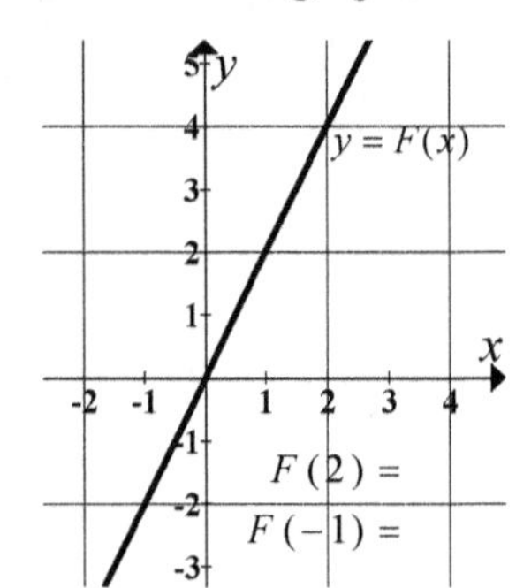

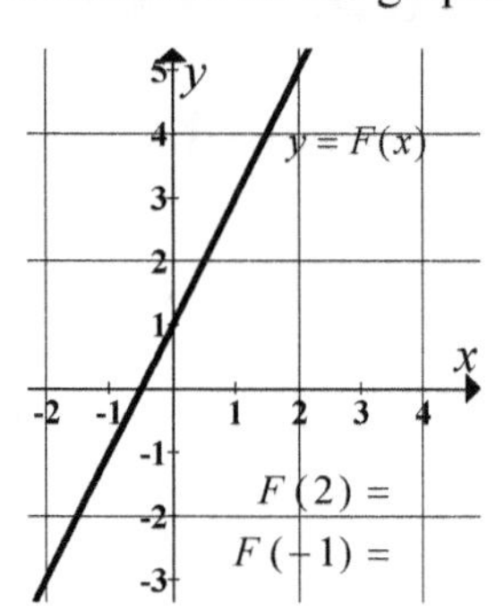

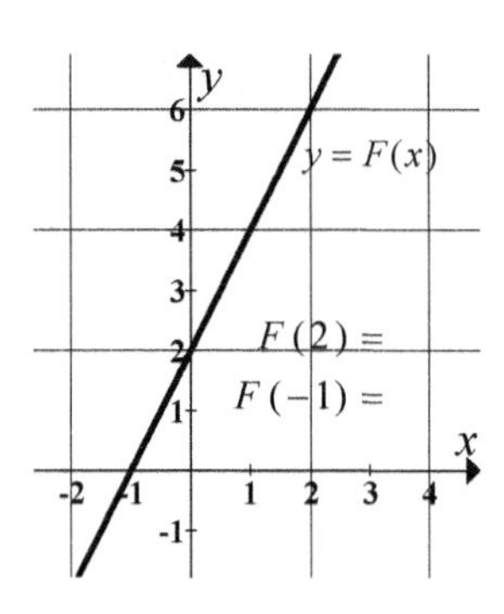

6. Graph ii in Model 1 shows the function $f(x)=2$. It turns out that *all* four functions (F) plotted in the previous question are antiderivatives of $f(x)=2$.

   a. Use the shaded area on Graph ii in Model 1 to confirm that $\int_{-1}^{2} 2\,dx = 6$. Then use the FTC to show that $\int_{-1}^{2} 2\,dx = F(2)-F(-1)$ for each antiderivative (F) regardless of which one you choose!

   b. Describe how the four antiderivatives of $f(x)=2$ at the top of this page differ.

   c. The equations of the functions plotted at the top of this page are shown below (listed from left to right). Find the derivative, F', for each function.

   $F(x)=2x-1$ $\quad$ $F(x)=2x$ $\quad$ $F(x)=2x+1$ $\quad$ $F(x)=2x+2$

   d. Give the equation of a function, not appearing in this activity, which is also an antiderivative of $f(x)=2$.

7. Read $F(2)$ and $F(-1)$ from each graph, and record these on the graph.

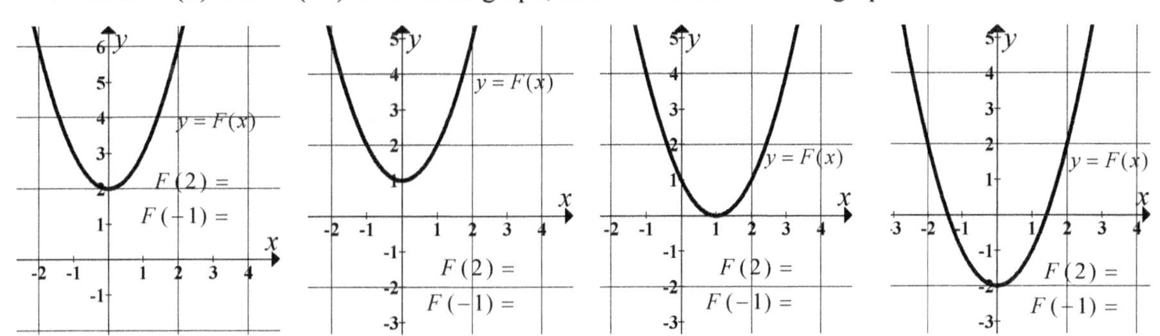

8. Use the shaded areas on Graph i in Model 1 to confirm that $\int_{-1}^{2} 2x\,dx = 3$.

 a. Which <u>one</u> of the functions F in the previous question is **<u>not</u>** an antiderivative of $f(x) = 2x$? Explain your reasoning.

 b. (Check your work) The equations of the functions plotted at the top of this page are listed below (from left to right).

 $F(x) = x^2 + 2$ $F(x) = x^2 + 1$ $F(x) = x^2 - 2x + 1$ $F(x) = x^2 - 2$

 Show how this information can be used to confirm your answer to part a.

 c. If C is a constant equal to any real number, what is $\frac{d}{dx}[x^2 + C] =$

9. Circle each that is true if $F(x)$ and $G(x)$ are antiderivatives of $f(x)$..

 $F(x) = G(x)$ $F'(x) = G'(x)$ $F(x) = G(x) + \text{a constant}$

Summary Box I6.1: Indefinite Integral

If f is continuous on $[a,b]$, and F is any antiderivative of f (that is, $F' = f$) …

then the **<u>indefinite integral</u>**, $\int f(x)\,dx$, designates the family of functions $F(x) + C$, where C is a constant with an infinite number of possible values (any real number). So we can write…

$$\int f(x)\,dx = F(x) + C$$

Note that this is different from the **definite integral** that we have been dealing with so far, which is written: $\int_a^b f(x)\,dx$, and is equal to the <u>number</u> $F(b) - F(a)$.

10. Give the **indefinite integral**, $\int f(x)\,dx$, for each function, f.
(Hint: Each answer should include a "C", as appears in Summary Box I6.1.)

a. $f(x) = 6x$ b. $f(x) = 6x^2$ c. $f(x) = 5x^4$

d. $f(x) = 3$ e. $f(x) = x^{17}$ f. $f(x) = 50x^9$

(Assume *k, r,* and *n* are constants, $r \neq 0, n \neq -1$)

g. $f(x) = k$ h. $f(x) = rx^{r-1}$ i. $f(x) = x^n$

Summary Box I6.2: Some Properties of Indefinite Integrals

Constant Multiples: $\int cf(x)\,dx = c\int f(x)\,dx$ (where c is a constant)

Sum (or Difference): $\int f(x) \pm g(x)\,dx = \int f(x)\,dx \pm \int g(x)\,dx$

11. Use Summary Box I6.2 to find ... (k and n are constants, $n \neq -1$)

a. $\int (15x^4 - 7)\,dx =$ b. $\int (12x^5 + 4x)\,dx =$ c. $\int 5e^x\,dx =$

Hint: recall that: $\frac{d}{dx}e^x = e^x$

12. Find the following derivatives...

a. $\frac{d}{dx}\left[\frac{x^{n+1}}{n+1} + C\right] =$ b. $\frac{d}{dx}[kx + C] =$ c. $\frac{d}{dx}\left[e^x + C\right] =$

13. (Check your work) Are your answers to the previous question consistent with the first row of Summary Box I6.3? Explain how the previous question is related to this first row.

14. Are your answers to Question 11 consistent with this first row of Summary Box I6.3? If not, go back and correct your work.

Summary Box I6.3: Indefinite Integral Formulas

If k and n are constants and $n \neq -1$ ...

$\int k\, dx = kx + C$ $\int x^n\, dx = \dfrac{x^{n+1}}{n+1} + C$ $\int e^x\, dx = e^x + C$

$\int \sin x\, dx =$ $\int \cos x\, dx =$ $\int \dfrac{1}{x}\, dx =$

(the following question will help you complete the bottom row)

15. Fill in each [ ] below with one of the following: $\ln x$, $\sin x$, $\cos x$, $-\sin x$, $-\cos x$, $\sqrt{x}$

$$\frac{d}{dx}\left[\qquad\right] = \frac{1}{x} \qquad \frac{d}{dx}\left[\qquad\right] = \sin x \qquad \frac{d}{dx}\left[\qquad\right] = \cos x$$

16. Use your answers to the previous question to complete the general antiderivative formulas in the bottom row of Summary Box I6.3.

17. (Check your work) For the last entry in Summary box I6.3 many students initially give the formula: $\int \frac{1}{x}\, dx = \ln x + C$. However, this works only when $x > 0$. For $x < 0$ one can show that $\frac{d}{dx}\ln|x| = \frac{1}{x}$. So the last formula in Summary Box I6.3 should be: $\int \frac{1}{x}\, dx = \ln|x| + C$.

   a. If necessary, correct your entry for the last formula in Summary Box I6.3.

   b. Find... i. $\int_1^e \frac{1}{x}\, dx =$ ii. $\int_{-e}^{-1} \frac{1}{x}\, dx =$ iii. $\int_{-3}^{-1} \frac{1}{x}\, dx =$

Extend Your Understanding Questions (to do in or out of class)

18. Find each of the following…

a. $\int (t+4)\, dt =$ b. $\int 3\sin x\, dx =$ c. $\int (5\theta - \cos\theta)\, d\theta =$

d. $\int \frac{1}{t}\, dt =$ e. $\int_0^{\pi} -\sin x\, dx =$ f. $\int_1^{3} (1-x^2)\, dx =$

g. $\int_1^{3} \frac{1}{x}\, dx =$ h. $\int \frac{1}{r^2}\, dr =$ i. $\int \left(\frac{1}{x^3}+\frac{1}{x}\right) dx =$

j. $\int_1^{4} 3\sqrt{x}\, dx =$ k. $\int_0^{\pi} (\cos x - \sin x)\, dx =$

19. (Check your work) In the previous question, identify each part that asks for…
    a. an indefinite integral and make sure your answer includes a constant C.
    b. a definite integral and check that your answer is a number.

20. For each function f, give the antiderivative, $F(x)$, for which $F(0) = -1$. Show your work.

a. $f(x) = -1$ b. $f(x) = 3x^2$ c. $f(x) = \sin x$

21. Explain why the formula $\int x^n\, dx = \frac{x^{n+1}}{n+1} + C$ does not work for $n = -1$. What is $\int x^{-1} dx$?

22. (Check your work) Check that your answer to the previous question is consistent with Summary Box I6.3.

23. Circle the letter of each pair of functions (v and s) for which $\int_0^a v(t)\,dt = s(a)$

a. $v(t) = 16t$, $s(t) = 8t^2 + 10$

b. $v(t) = t$, $s(t) = 0.5t^2 - 1$

c. $v(t) = -4t$, $s(t) = -2t^2$

d. $v(t) = 0.25t$, $s(t) = 0.125t^2$

Show your work and explain your reasoning.

24. The position of an object under the influence of a constant force and with initial velocity 0 (e.g., a freely falling object dropped from rest) can be expressed as $s(t) = k \cdot t^2 + C$, where the magnitude and sign of k gives the strength and direction of the constant force, and t is time. Each position function in the previous question can be expressed in this form.

a. For each function s that you **did not** circle in the previous question, what is the value of $s(0)$? Write this value (i.e., $s(0)$ = <*value*>) below each pair.

b. For each function s that you circled in the previous question, what is the value of C? Write C = <*value*> below each pair.

c. What does C in the formula $s(t) = k \cdot t^2 + C$ tell you about the object? Explain your reasoning.

d. (Check your work) Are your answers to parts a-c consistent with the fact that $\int_0^a v(t)\,dt = s(a)$ only if the initial position of the object is zero? Explain.

Notes

Printed in the USA
K061016SCI081117 01S29053000000002135